D1126594

NO WORLD WITHOUT END

The New Threats to Our Biosphere

NO WORLD

The New Threats

by KATHERINE

WITHOUT END

to Our Biosphere

AND PETER MONTAGUE

OCEAN COUNTY COLLEGE
LEARNING RESOURCES CENTER
TOMS RIVER, N.J. 08753

G. P. Putnam's Sons, New York

574.52
M 759 N

Copyright © 1976 by Southwest Research and Information
Center

All rights reserved. This book, or parts thereof,
may not be reproduced in any form without permission.
Published simultaneously in Canada
by Longman Canada Limited, Toronto.

SBN: 399–11500–5

Library of Congress Catalog
Card Number: 74-30569

PRINTED IN THE UNITED STATES OF AMERICA

To our families (parents, siblings, and children) who have borne us and borne with us; and to C. H. Lumpkin and to Mary D. Ragosin, who at important times guided us; but especially to Jessica, Timothy, and Tyree

76573

CONTENTS

Acknowledgments

WE would like to thank some of the people who in one way or another helped us during the long period while we were writing this book. Most obviously, our parents (to whom we dedicated our first book, *Mercury*, though the publisher omitted the dedication by mistake).

Only a little less obvious are all of our colleagues past and present at the Southwest Research and Information Center (Albuquerque, New Mexico), as well as the center's supporters. The center is a problem-focused research organization; its clipping files and source books are an extraordinarily useful information resource available to the general public. A book like this, broad in scope, is only one of the possible uses to which the center's files can be put.

Sandra Simons, formerly a researcher at the center, deserves special thanks for her tireless after-hours work researching and typing an early draft. She has been a constant friend, as has Kenneth Schultz, who is now the center's senior research associate. Charles Hyder, staff scientist at the center, and close friend, criticized early drafts of the manuscript and made major contributions to the book and to our conception of the way things are.

Sheldon Novick, Kevin Shea and Julian McCaull, who edit *Environment* magazine, criticized an early draft of the manuscript. Our understanding of agriculture and of many aspects of the energy industry owes much to the enormous wealth of information which *Environment* has published during the past fifteen years.

It was Sheldon who first made us see, in the mid-1960s, that the environment was the most important "issue" facing humankind and that the resolution of the world's critical environmental problems in the next thirty years is going to determine all human relations in the future. Through Sheldon and Kevin (who are members of the Board of Directors of Southwest Research and Information Center) and through *Environment,* we owe a major intellectual debt to Dr. Barry Commoner and his colleagues in the "scientific information movement" which they created in the late 1950s and early 1960s. Much of the present-day "public interest research movement" has developed from their original Committee for Nuclear Information in St. Louis.

Michael Jacobson and Al Fritsch of the Center for Science in the Public Interest (Washington, D.C.) each criticized parts of the early manuscript and widened our thinking about food and about general problems of exotic chemical substances loosed into the biosphere.

None of these people, of course, bears any responsibility for any of the specific content of the book; that responsibility rests with us.

There are other people to whom we are intellectually indebted, people who have inspired thought and action in various ways. Aileen Adams, Geoff Cowan, Garrett De Bell, Anne and Paul Ehrlich, W. H. Ferry, Ray A. Graham III, Robert Harris, Harvey Mudd II, Glenn Paulson, Donald Ross, Henry Schroeder, Robert Scrivner, Irving Selikoff, and James Sullivan are among the most important. We owe special thanks to Ralph Nader, whose work has energized the public interest research movement for a decade.

We are grateful to the staffs of the Library of Medical Sciences and the Zimmerman Library at the University of New Mexico, as well as to the staff of the Albuquerque Public Library. Their consistently helpful work has been invaluable. The same is certainly true of our editor, Edward Chase, and his assistant, Charles Finberg, who prepared the manuscript for printing and helped shape the book in the process. Our thanks to Dan Butler, who typeset the tables for us.

Don Schlegel, former chairman of the Architecture Department, and Joel Jones, former director of American studies—both at the University of New Mexico—deserve special note for the effective academic environments they have created for interdisciplinary research. We frequently read that interdisciplinary study is essential to the solution of critical problems threatening the future of all humankind. Schlegel and Jones in their own ways made the abstract "interdisciplinary study" a working reality, and they managed it within the context of a poorly funded state university.

Finally, our special thanks to Marie Rodell, our kind and thoughtful literary agent and friend.

KATHERINE AND PETER MONTAGUE

Albuquerque, New Mexico
July 4, 1975

Introduction

"It seems obvious . . ."

WE wrote this book to help people understand important facts about the physical basis of industrial civilization and its attendant dangers. The book describes many new threats to our biosphere and to ourselves. It concludes with numerous concrete suggestions for change, for improvement.

As the reader will see, this is a book about the human body and about the earth, and it is a book about many new ways in which people are threatening to bring an end to life as we know it. The book is intended to help readers understand that there can be a physical and chemical basis for guiding necessary political and ethical action.

The earth consists of ninety-two naturally occurring chemical elements. Of these, seventy-two are metals. In simplest terms, metals can be defined as "elements generally characterized by ductility, malleability, luster, and conductance of heat and electricity."[1] All the ninety-two elements have unique characteristics which are determined by the makeup of their different atoms. These unique characteristics suit each element for different purposes in nature, and they also make at least fifty-two metallic elements industrially useful. For example, powdered aluminum was recently developed into the explosive material for America's largest nonnuclear bomb, the 7.5-ton "Daisy Cutter."[2]

Lead serves as a lubricant inside automobile engines, and it also improves the burning qualities of gasoline, which is why Americans started putting thousands of tons of toxic lead into their gas tanks in 1923, a practice we are only beginning to abate today.[3] Starting more than 120 years ago, the addition of carbon, molybdenum, chromium and nickel turned large quantities of soft iron into hard steel and thus made possible modern industrial life.[4]

We live in the age of metals, and contemporary American civilization would be unthinkable without an abundance of cheap metals. Aluminum, iron, chromium, copper, lead, zinc, tin, mercury, and forty-four other metals serve as the foundation blocks of all our industrial achievements. Our chemical industries (rubber, plastics, dyes, synthetic textiles), our energy production, our transportation systems, our high-yield agriculture—in short, the material bases of our national wealth—all depend critically on the large-scale mining and refining of metals or metal-bearing minerals. It seemed relevant, therefore, to write a book about metals and the environment, hoping to discover something about America in the process.

What we discovered is that our civilization is characterized by a strong sense of apocalyptic doom and impending disaster. Of course, this could be viewed as simply an extension of a long American tradition of apocalyptic writing. We recognize that America has a long history of unwarranted doomsaying, but we also know that better than 99.9 percent of the apocalyptic tradition is based on an expectation of supernatural events. Moreover, from reading many of the important policy documents published by Establishment institutions during the past decade, we gather that modern apocalyptic writers differ significantly from their tradition in this fundamental respect: Modern doomsayers are arguing from a basis in physical fact. Their conclusions are not based on some private chiliastic pipeline to heaven and the Truth. Their conclusions are based on observation and on reasoning, on verifiable information open to inspection by all.

In addition writers in the new apocalyptic wave are commonly not just authors but are scientists by profession. For example, John Platt, a research biophysicist, writing in *Science,* the journal of the American Association for the Advancement of Science, in November, 1969, said:

There is only one crisis in the world. It is the crisis of transformation. The trouble is that it is now coming upon us as a storm of crisis problems from every direction. But if we look quantitatively at the course of our changes in this century, we can see immediately why the problems are building up so rapidly at this time, and we will see that it has now become urgent for us to mobilize all our intelligence to solve these problems if we are to keep from killing ourselves in the next few years.

Whether we have 10 years or more like 20 or 30, unless we systematically find new large-scale solutions, we are in the gravest danger of destroying our society, our world, and ourselves in any of a number of different ways well before the end of this century.[5]

The well-known American botanist/ecologist Barry Commoner said in the April, 1970, issue of *Environment*:

We are in an environmental crisis which threatens the survival of this nation, and of the world as a suitable place for human habitation. Environmental pollution is not to be regarded as an unfortunate but incidental by-product of the growth of population, the intensification of production, or of technological progress. It is rather an intrinsic feature of the very technology we have developed to enhance productivity. Our technology is enormously successful in producing material goods, but too often is disastrously incompatible with natural environmental systems. . . .

Unless we start now with a *fundamental* attack on the environmental crisis, we will find ourselves, in a decade, locked into an irreversible, self-destructive course.[6]

Dr. Commoner has repeatedly emphasized this message:

I believe that we have, as of now, a single decade in which to design the fundamental changes in technology that we must put into effect in the 1980s—if we are to survive.[7]

To establish the physical basis for understanding America's situation today, we begin with a brief history of the earth and the

human body. Next we discuss the immense modern event which
has been called the chemicalization of the environment. Then we
move to case studies of particularly troublesome and persistent
chemical elements, the toxic heavy metals cadmium, lead, and
mercury.

At that point the focus of the book moves outward to
metals-related problems. We briefly discuss the nation's two most
extensive environmental health problems—the workplace and the
inner city. Then we turn from the city to the countryside where we
describe the current agriculture situation, which has been described elsewhere as an ecological disaster area. Modern agricultural practices lead to a discussion of essential trace elements (many of
them metals) missing from the American diet. From there we move
to the general state of health of the American people, where there
is evidence of an increase in disease and a decrease in longevity.
Next, we look at the global-sized problems of environmental deterioration, stressing the dangers of simplifying and destabilizing the
planet's ecosystems. Finally, after focusing on the imminent world
shortages of land, metals, and safe sources of energy, we discuss, in
a summary chapter of recommendations, some of the changes necessary.

As John Platt has observed, we are living in a unique historic
period, a century during which our speeds of communication have
increased by a factor of 10,000,000; our speeds of travel by 100;
our speeds of data handling by 1,000,000; our ability to control
diseases by something like 100; our rate of population growth by
1,000 times what it was just a few thousand years ago.[8]

The intense changes taking place in our time will probably not
recur. Our communications will probably never get much faster
than they are now; our bombs will not get much bigger; our per
capita use of energy may double or triple, but after that it will most
likely peak and then stabilize or even diminish. If we can survive
this period of intense change, there is a reasonable possibility that
we can emerge into a world of dynamically stable relationships
between people and people and between people and the earth. We
could conceivably achieve moderate abundance for all, and
perhaps even worldwide peace.

The imperative moral question that has been thrust upon our
generation of Americans to answer is: Can we make it through the

five- to thirty-year term without bringing on the extinction of our species and without destroying the biosphere?

For more than a decade many well-known scientists and scholars, such as Harrison Brown, Georg Borgstrom, Rachel Carson, Lester Brown, Barry Commoner, René Dubos, Anne and Paul Ehrlich, Samuel Epstein, G. Evelyn Hutchinson, Gunnar Myrdal, Leo Szilard, U Thant, George Wald, and hundreds of less well-known people have tried to alert us and arouse us to imminent crisis conditions building in our world. Because of their alarums, we could not write a book narrowly about metals. There is an obvious need at this time for writers to attempt interdisciplinary, synthesizing studies that try to make sense out of a troubled world. We hope this book goes a step in that direction.

> It seems obvious that before the end of the century we must accomplish basic changes in our relations with ourselves and with nature. If this is to be done we must begin now. A change system with a time lag of ten years can be disastrously ineffectual in a growth system that doubles in less than fifteen years.[9]
> —STUDY OF CRITICAL ENVIRONMENTAL PROBLEMS, 1970

1

Where We Came From

History of the Earth and the Human Body

IN the beginning, more than 5 billion years ago, cold clouds of dust and gas grew into a whirl of nuclear "fire." Eventually the fire settled into a cyclical pattern we call our solar system. As the radioactive fire cooled, perhaps 4.5 billion years ago, the earth took solid form.[1] In its early years the earth was a very different place from what we know today. The atmosphere didn't contain free oxygen but was probably a smelly mixture of water vapor, carbon dioxide, methane gas, and ammonia (though, of course, there was no one around to smell it). The earthly environment, in those days before life evolved, was mostly stark radioactive rocks, thick mist, intense ultraviolet radiation with the sunlight, and little else. The earth wouldn't take on its modern blue-green color until the biosphere developed, over billions of years.

As the earth cooled and water vapor condensed out of the atmosphere, oceans formed. Slowly the early oceans began to dissolve and mix together the 92 chemical elements from the earth's crust. For more than a billion years the lifeless oceans existed, weathering and leaching chemicals out of the earth's rocky crust. Eventually large portions of the planet lay covered by a complicated chemical mix which has been called protoplasmic soup. It was here in this vast, steamy soup kitchen that the first sign of life, the first simple amino acid, happened into being. The first amino acid was a combination of four elements: carbon, hydrogen, oxygen, and nitrogen. In this earliest detectable ancestor of modern life, we can observe one of the fundamental chemical principles of biology: All life is carbon-based.[2] The unique atomic structure of the carbon atom allows it to form many strong chemical bonds with other elements. Carbon is so important in the earth's biochemistry that scientists recognize two major classes of chemical compounds: those which contain carbon (called organic compounds) and those which contain no carbon (the inorganic compounds). The carbon-based (organic) compounds grow much larger and more complex than the inorganic compounds. All living (organic) molecules are huge compared to, say, a molecule of water. A molecule of water (an inorganic substance) has an empirical formula which is written H_2O, meaning that the molecule is constructed of two atoms of hydrogen and one atom of oxygen. By way of contrast, a typical molecule of animal protein (in this case, codfish protein—a typical organic molecule) has an empirical formula which is written $H_{555}C_{265}O_{174}N_{83}S$, indicating 555 atoms of hydrogen, a single atom of sulfur, and amounts of nitrogen, oxygen, and carbon in between. It is principally the carbon atom that makes this complexity possible, though of course it doesn't manage the job alone.[3]

No one really knows precisely what evolutionary steps led to the development of living things. But contemporary scientists have developed plausible theories based on evidence and on reasoning. According to these theories, after the first amino acids evolved, larger structures became possible—structures linking amino acid to amino acid in long chains called proteins. To help form the necessary chemical links, the protoplasmic ocean soup provided metals—zinc and magnesium and others—which did the job very well. Today zinc is still used by all living organisms to manufacture proteins.[4] From the earliest beginnings of the planet Earth, metals have played a fundamental part in the basic chemistry of life.[5]

As the oceanic soup mixed and thickened, larger and more complex proteins evolved until one time, a little more than 3 billion years ago, at a place near present-day South Africa, an organism developed which was large enough to leave a first fossil record in the rocks (the so-called Figtree fossils).[6] This early creature couldn't manufacture its own food but had to depend on bumping into nutrients floating in the soup, an inefficient process at best. However, clear fossil records show, after another billion years of evolution, the existence of organisms that could make their own food out of water and carbon dioxide, using sunlight for energy in a chemical process called photosynthesis. Photosynthesis was an enormously influential biological development. Photosynthesis is undoubtedly one of the most important biological reactions on earth, because ultimately all life is dependent on solar energy captured by photosynthesis. The photosynthetic ability made new food sources widely available: carbon dioxide and water, which were abundant in the oceans. As a result, photosynthetic creatures would eventually take over the planet, changing it radically as they went, creating the biosphere—the thin film of life forms that covers the earth, the environment as we know it today.

Here again we see the early importance of metals in various life forms. The metal magnesium makes photosynthesis possible. One atom of magnesium metal lies at the center of each molecule of the photosynthetic compound chlorophyll; chlorophyll is the greenish-yellow compound which gives green plants their color. Chlorophyll has an empirical formula written $C_{55}H_{72}O_5N_4Mg$ (Mg being the chemical symbol for magnesium). Chlorophyll makes possible one very fundamental reaction in plants: it allows plants to convert water and carbon dioxide into sugars and starches, thus capturing the energy of sunlight and storing it away.[7] These sugars and starches (carbohydrates) are energy-rich compounds, high in calories, which animals can digest (metabolize) and use to make the substance of their own body tissues. At its simplest, the photosynthetic process can be shown this way, using empirical formulae: water (H_2O) plus carbon dioxide (CO_2) plus sunlight gives you the simple carbohydrate formaldehyde (HCHO), plus free oxygen (O_2) plus stored energy from the sunlight.[8] Using this basic formula, plants manufacture complex carbohydrates (sugars, starches) and thus provide the world with its primary food supply. For this reason, photosynthetic plants are called primary producers.

In addition to its main products (sugars), photosynthesis has an equally important by-product which plants excrete into the air: free oxygen. Almost all the free oxygen in the earth's atmosphere is believed to have originated from photosynthesis (though some has been created by the dissociation of water). The Russian-born American biochemist Eugene Rabinowitch, a leading investigator of photosynthesis, has estimated that each year the earth's green plants combine a total of 150 billion tons of carbon (from carbon dioxide) with 225 billion tons of water, excreting (or liberating, if you will) 400 million tons of oxygen.[9] Stated another way, each year photosynthesis produces a new plant mass of vegetation which weighs between 150 and 200 billion tons (dry organic matter) and which includes both food for people and the energy that runs all the life-support systems of the biosphere, namely the earth's major ecosystems: the oceans, forests, grasslands, marshes, estuaries, lakes, rivers, tundras, and deserts.[10]

This is a very important point to grasp. Photosynthesis is the chief means whereby energy gets into all of the earth's ecosystems, and all ecosystems require constant energy inputs or they stop working. Not only do photosynthetic plants provide the atmosphere with oxygen, but they also provide all life forms with the energy (calories) to live and grow. As Yale biologist G. Evelyn Hutchinson has stressed:

> . . . The energy of solar radiation can enter the biological cycle only through the photosynthetic production of organic matter by chlorophyll-bearing organisms, namely green and purple bacteria, blue-green algae, phytoplankton and the vast population of higher plants.[11]

Hutchinson goes on to calculate the thickness of chlorophyll-bearing life in a typical lake; it totals 100 microns (100μ) or less, which is about the thickness of this page. In the ocean, Hutchinson calculates, the layer of chlorophyll-bearing life is even thinner. In a forest, the process is probably even still less efficient and less substantial. It is necessary, Hutchinson says

> to maintain a balance in our attitude by stressing the fragility and inefficiency of the entire process. . . . The machinery by which energy enters the living world is clearly quite tenuous.[12]

It is worth pointing out that there exists a chemical similarity between chlorophyll and hemoglobin (the iron-containing part of human blood).[13] In both these important protein substances, a metal lies near the molecular center. If for any reason the metal is displaced, the chlorophyll (or the hemoglobin) stops working. In both cases, the metals make possible the same basic job—they help form proteins which hook onto oxygen and carry the oxygen around and then drop it off where it is needed (where its partial pressure is low). As we shall see, oxygen is a critical substance in modern life forms, and the metals that help carry it around make modern life forms possible.

It was oxygen that precipitated the first worldwide ecological crisis. It happened between 2 and 3 billion years ago, when organisms first developed the photosynthetic ability and began to excrete pure oxygen. Up to that time probably no free oxygen existed on earth; it was probably all tied up in carbon dioxide (CO_2), in water (H_2O) or in other basic inorganic chemical compounds. Pure oxygen is a very reactive chemical—it hooks on (via "burning") to almost anything that it touches, forming a new compound. Thus free oxygen was a powerful poison to all life forms that were getting along before photosynthesis developed. Because photosynthetic creatures could manufacture their own food from water and carbon dioxide, they could grow much faster than their ancestors, and they spread throughout the oceans, using the energy of sunlight to stoke the chemical fires of their own growth, spewing poisonous oxygen behind. Biologist G. Evelyn Hutchinson has characterized the spread of photosynthetic (oxygen-excreting) creatures as "certainly the greatest biological revolution that has occurred on the earth. The net effect of this revolution was no doubt the extermination of a great number of inefficient and primitive organisms that could not tolerate free oxygen and their replacement by more efficient respiring forms."[14]

The triumph of the photosynthetic creatures came with the development of a basic chemical mechanism for handling free oxygen. This basic mechanism goes by the name of enzyme, and with its development we enter the modern period of the history of life— the age of enzymes (though this period is still about 2 billion years in the past). More on enzymes later.

In the modern period the blue-green biosphere as we know it evolved at an increasingly rapid rate. The seas thickened to their present salty mixture as the earth's original radioactivity slowly subsided. Oxygen, carbon, hydrogen, and nitrogen slowly established themselves in their present proportions in air, water, and soil and began undergoing the enormous self-regulating cyclical *living* events which now maintain the earthly environment in dynamic equilibrium. Ecologist Barry Commoner has recently described these great cycles of the biosphere:

> All living things, including man, and all human activities on the surface of the earth, including all of our technology, industry and agriculture, are dependent on the great interwoven cyclical processes [which are] followed by the four elements that make up the major portion of living things and the environment: carbon, oxygen, hydrogen and nitrogen. All of these cycles are driven by the action of living things: green plants convert carbon dioxide into food, fiber and fuel; at the same time they produce oxygen. . . . Plants also convert inorganic nutrients [carbon dioxide, nitrates, and phosphates] into foodstuffs [for themselves and for animals]. Animals, basically, live on plant-produced food; in turn they regenerate the inorganic materials—carbon dioxide, nitrates and phosphates—which must support plant life. Also involved are myriads of microorganisms mainly for decomposition of organic matter in the soil and water. Altogether this vast web of biological interactions generates the very physical system in which we live: the soil and the air. It maintains the purity of surface waters and, by governing the movement of water in the soil and its evaporation into the air, regulates the weather. This is the environment. It is a place created by living things, maintained by living things, and through the marvelous reciprocities of biological evolution is essential to the support of living things.[15]

As life forms grew larger and more complex, they achieved the miraculous structure (and qualities) of the cell. According to fossil records, this apparently happened between 1.2 and 1.4 billion years ago.[16] The cell, when it appeared in the oceans, consisted of protoplasmic soup inside a little membrane or bag. This device still serves as the basic building block for all forms of life today. The human body, for example, is a collection of 50 trillion small bags of protoplasmic soup (cells) all arranged together in just the right relationship to keep everything balanced and stable and functioning.

In spite of the great present diversity of living things, all share the same basic ground plan. Their cells all carry out metabolism in pretty much the same way.[17] "Metabolism" is the collective name for all the thousands of chemical processes an organism goes through when it takes in some food, breaks down the food into its component parts, distributes the parts throughout its body, and builds new tissue for itself, excreting any materials that were useless or poisonous in the food.

All 50 trillion cells in the human body continually carry out the many steps which go on under the name of metabolism; it is a twenty-four-hour-a-day process. Here, in the fundamental cell chemistry of the body, we again see the importance of trace metals. One crucial factor in most body processes is a kind of protein called an enzyme. Proteins are large organic molecules made up of amino acids strung together like beads on a string in different combinations. Enzymes are proteins which act like keys; they unlock chemical compounds, rapidly and precisely breaking food into its constituent parts. Enzymes also serve to get energy into cells, in the fundamental process called cell respiration.[18] All living cells require a continuous supply of energy to remain alive, so it is clear that enzymatic cell respiration is one of the most fundamental processes in biology.

Enzymes have a remarkable ability to reduce the amount of energy required to get a particular chemical reaction going. The Scottish researcher Dr. J. M. A. Lenihan has said:

> A chemist who tries to decompose a protein in the laboratory would probably have to boil it in concentrated hydrochloric acid for 24 hours; yet this process is going on throughout the body at a much faster rate and without benefit of high temperatures or corrosive reagents. These achievements are the result of catalysis by enzyme activity.[19]

Enzymes carry out the crucial chemical function called catalysis. Acting as catalysts, enzymes initiate or change the speed of chemical processes to a tremendous degree, usually speeding them up. Catalysts do not permanently participate in the chemical reactions which they catalyze; they simply enter a situation, promote a reaction between two or more other chemicals, then step back out of the reaction and move on to perform their job again elsewhere. For example, one molecule of the enzyme catalase, the biological catalyst that decomposes hydrogen peroxide, can promote the decomposition of 40,000 molecules of hydrogen peroxide per second at freezing point.[20] The human body contains trillions upon trillions of enzymes; each individual cell may contain several thousand.

At least 700 different enzymes have been identified in animals, plants, and microorganisms.[21] Of these, at least 200 are known to depend on a metal for activity. There are thousands of other enzymes which have never been taken apart by analytical chemists and which thus remain "unidentified" though it is known that they *are* enzymes by their large molecular size and by the way they act.[22]

When we try to estimate the possible number of different enzymes, the number must be very large because there are at least 2.4×10^{18} (2.4 quintillion) possible different proteins, and most enzymes are proteins with something extra added on. The something extra is called a coenzyme, and it can be an organic molecule (a vitamin) or a metal or a vitamin that *contains* a metal (like vitamin B_{12} which contains the metal cobalt and which ceases functioning as a coenzyme if deprived of its cobalt).[23]

Many enzymes depend entirely on a metal ion (or atom) being attached to their protein portion. An ion is simply an atom or a molecule carrying an electrical charge (positive or negative). If the proper metal isn't available—as in the case of a dietary deficiency—the enzyme will not be active. On the other hand, not only deficiencies but also excesses of a metal can deactivate an enzyme. In the words of J. E. Kench, "Around all the macromolecular structures in the living cell hovers a cloud of metal ions, jostling for position on the surface of the large molecules and according to their numbers and characters, in the case of enzymes, helping or hindering the movement of molecules . . . to and fro."[24] Different metals have differing attractiveness to different enzymes. In addition, enzymes are extremely specific—they must have one particular metal, in most cases, or they deactivate.

Thus, an excess of one metal can displace the proper metal from the surface of an enzyme protein in a process called competitive inhibition and knock the enzyme out just as a deficiency would. An excess of one element induces a deficiency of another. Here we can begin to appreciate the idea of a balance of nutrients. Nutritional excesses and deficiencies are all relative.

When one metal displaces another metal on a coenzyme and thus knocks out the enzyme itself, we are seeing the action of an enzyme poison. Enzyme poisons are particularly complex and difficult to understand because they may attack so many different parts of a living system (a plant or an insect or a person, for example). The metals arsenic, lead, mercury, and cadmium function as enzyme poisons in people and other life forms, as we shall see.

Whether in air or in water or inside the human body, the sizes of particles, ions, and cells are significant; frequently the precise chemical and medical effects of a substance will depend on the size of the particles in which the substance appears (and on their electrical charge).

Dissolved noncolloidal materials, such as metal ions (which are simply metal atoms carrying a positive or negative electrical charge) are extremely small, measuring 0.005 microns or less in diameter. Fifty million metal ions laid next to each other would equal one inch. By comparison, proteins are huge and cells are gigantic. The human ovum—the single cell found in females which, when fertilized by a sperm cell from a male, starts multiplying to produce a new human being—measures about 100 microns in diameter. This large cell can just barely be seen by the human eye unassisted by microscope. A more normal-sized human cell measures about ten microns in diameter and is invisible without magnification.

By way of contrast, bacteria (small forms of single-celled life which, among other things, can cause disease in humans, animals, and plants) measure about 0.4 microns in diameter. Viruses, on the other hand, are very much smaller than bacteria, averaging perhaps 1/50th of a micron (0.02μ). Viruses enter cells and bacteria do not.

Whether or not a particular substance enters a human cell depends on the size of the substance and on its electrical charge. Human cells appear in immense varietes of size and shape and chemical structure (though they all share certain common operating features). Physically, some cells are long and strung out and wispy while others are very short and compact, almost round. Inorganic chemicals (for example, metal ions, swallowed by a person as a drinking water contaminant) will hover around the cells, competing for positions near the cell wall (properly called the cell membrane). Here, depending on the size and electrical charge of the substance, the metal ion can become an active part of the cell wall itself or can pass right through into the soupy interior. Once inside the cell, swimming about in a tremendously complex electrochemical environment, the metal ion will behave in any of dozens of ways, helping or harming the cell's activities, depending on whether cells have evolved ways of handling the particular ion. As is obvious, life evolved in a certain chemical environment that changed very slowly, and cells achieved great complexity of structure during their billions of years of evolution. Any abrupt introduction of new elements (significant changes over decades or less)—such as mercury from eating a lot of swordfish—brings about destructive changes in cells, impairing life's capacity to go on.

Of the presently identified 700 enzymes, some 200 (roughly 30 percent) depend on the proper metal being attached to them as a coenzyme or activator.[25] Dr. Henry Schroeder has published an incomplete list of the so-called metalloenzymes; they have names like glycyglycine dipeptidase, so we won't reproduce the list. We should just note that seven major classes of metalloenzymes are known and that at least the following metals have been identified with healthy enzyme activity: copper, manganese, zinc, magnesium, iron, cobalt, molybdenum, and vanadium. As Dr. Schroeder stressed when he published his list of metalloenzymes, "the list is by no means complete, and undoubtedly will be expanded in the future as enzymes are purified."[26]

Enzymes continually carry out essential operations in both plants and animals; here is one example of the many ways in which they can be important: Plants absorb their nitrogen from the soil in the form of a nitrate, but they use it in their leaves in a different form. To change the chemical form of the nitrate, an enzyme called nitrate reductase goes into action. A molecule of nitrate reductase enzyme must always contain an atom of the metal molybdenum. If molybdenum is not present, the nitrate reductase becomes inactivated, the nitrates in a plant cannot change their chemical form, and the plant starves despite adequate nitrates in the soil. Only two ounces of molybdenum per acre of land (138 grams per hectare) will keep nitrate reductase activity at healthy levels. If soil contains too little molybdenum, the nitrate level can get so high in plants that the plants become poisonous to cattle. When this kind of "secondary poisoning" effect became understood—not over thirty years ago—one of the problems was to design fertilizer equipment that could disperse such small amounts as 2 ounces per acre of a nutrient. As is obvious, an increase in the use of a nitrate fertilizer automatically increases a crop's need for molybdenum; if additional molybdenum is not supplied (either from the soil itself or as an additive), the balance of nutrients can be upset with far-reaching consequences.

Although enzymes are by far the most important group of catalysts in the human body (indeed, in all living organisms, since even tiny bacteria depend on enzymes to carry out metabolism), there are other catalysts that are not enzymes. Metals by themselves, not attached to proteins, can act as catalysts in many chemical reactions. One example is the metal boron, which is essential to plant growth. Without participating in an enzyme, boron catalyzes an important reaction which removes sugars from a plant's leaves after photosynthesis. If boron is not present, a plant's leaves begin to fill with sugars and the plant will die.[27]

It should be obvious from this very brief discussion of cells and their extremely complex dependence on enzymes (and the frequent dependence of enzymes on metals) that metals are of critical importance in human metabolism and the metabolism of all living organisms. Both excesses and deficiencies can be debilitating and lethal.

Let us resume our history of the human body. Once the cell had developed, it reproduced (by splitting itself into two cells, then four, then eight . . .) and spread throughout the oceans. As the volume of free oxygen in the atmosphere grew to be substantial, a band of ozone (with the empirical formula O_3) formed high above the earth (12 to 25 miles up in the sky). This band of ozone filtered out most of the sun's deadly ultraviolet rays, making it safe for life to emerge from the protective covering of the oceans.[28] A spurt of growth resulted; life multiplied and grew increasingly complex and diverse and therefore stable and enduring. (Today, as we shall see, human activity is threatening to destroy the earth's protective ozone layer.)

The first fossil animal skeletons were laid down in ocean sediments some 600 million years ago.[29] After that it took another 200 million years before any organism left the ocean and invaded the land. First to make the break were the plants. The plants made it possible for modern soil to develop. After the plants got themselves

firmly established on land (in probably just a few million years),
then the animals could follow. The first animals to make the move
were the arthropods (crabs, lobsters, insects), the mollusks (oysters,
snails, slugs), and the worms. [30]

At this point there was a brief 90-million-year wait while the first
mammals evolved. During this period, creatures were adapting to
their new circumstances. The nutritional needs of the earliest land
dwellers had been established during very long periods of ocean
and then freshwater living. In their cells, the new land creatures
(both plants and animals) still carried a salty solution that was very
much like seawater, and they had to maintain that salty solution or
die. Tables 1-1 through 1-4 show some of the important chemical
relationships between the solar system, the earth, seawater, and living things. Creatures had to maintain a balance of the elements to
keep their cells working. This is still what nutrition is all about—
keeping the balance of nutrients in the right ranges for the needs
of your cells.

The new land environment, then as now, contained

> . . . an abundance of sodium, potassium, magnesium,
> calcium, and a super-abundance of silicon, aluminum,
> iron, titanium, cobalt, manganese, chromium, fluorine,
> rubidium, and barium. It was partly or relatively deficient
> in chlorine, bromine, iodine, selenium, molybdenum and
> boron. . . . Sodium was not evenly distributed.
> . . . Whereas strontium, vanadium, chromium, man-
> ganese, iron, copper and zinc were fairly ubiquitous, they
> varied . . . in concentrations from place to place,
> depending upon solubility and leaching by rainfall. Co-
> balt, boron and molybdenum were not distributed
> uniformly.[31]

To keep their cells supplied with essential minerals, the plants
and animals developed homeostatic mechanisms. Homeostasis
(from Latin and Greek words meaning roughly "stay the same") is
what keeps our bodies from containing too much copper, for example, because large amounts of copper are poisonous to cells.[32]

Copper is an essential trace element but too much of it is toxic.
This is true in the case of just about everything except air. Too
much table salt can kill you—so you have homeostatic mechanisms

Table 1-1. Some chemical/physical realities: chemical composition of the universe, the earth's crust, seawater and the human body.
(Numbers inside parentheses indicate percentage of total number of atoms.) See also Tables 1-2 and 1-3, which give data on trace rather than bulk elements. (Note that data in Tables 1-2 and 1-3 differ slightly from data given here, indicating rapidly changing analytic techniques in the biosciences.)

Composition of the universe (in percents)	Composition of earth's crust (in percents)	Composition of seawater (in percents)	Composition of human body (in percents)
Hydrogen (91)	Oxygen (47)	Hydrogen (66)	Hydrogen (63)
Helium (9.1) *	Silicon (28)	Oxygen (33)	Oxygen (25.5)
Oxygen (0.057)	Aluminum (7.9)	Chlorine (0.33)	Carbon (9.5)
Nitrogen (0.42)	Iron (4.5)	Sodium (0.28)	Nitrogen (1.4)
Carbon (0.021)	Calcium (3.5)	Magnesium (0.033)	Calcium (0.31)
Silicon (0.003)	Sodium (2.5)	Sulfur (0.017)	Phosphorus (0.22)
Neon (0.003)	Potassium (2.5)	Calcium (0.006)	Chlorine (0.03)
Magnesium (0.002)	Magnesium (2.2)	Potassium (0.006)	Potassium (0.06)
Iron (0.002)	Titanium (0.46)	Carbon (0.0014)	Sulfur (0.05)
Sulfur (0.001)	Hydrogen (0.22)	Bromine (0.0005)	Sodium (0.03)
All others (less than 0.01)	Carbon (0.19)	All others (less than 0.1)	Magnesium (0.01)
	All others (less than 0.1)		All others (less than 0.01)

* Totals do not come to exactly 100 percent because of rounding.

Table 1-1 adapted from Earl Frieden, "The Chemical Elements of Life," **Scientific American,** Vol. CCXXVII (July, 1972), p. 54.

---

to keep you from taking in too much salt. If you do take in too much, your body promptly tries to get rid of it by various methods (sweat, vomit, urine, for example).

On the other hand, the body's homeostatic mechanisms will act to conserve salt in the face of a salt-deprived diet. There are various homeostatic mechanisms in our bodies and these mechanisms are frequently associated with our detoxification mechanisms—the ways in which our bodies get rid of poisons when we accidentally ingest

Table 1-2. Major (or Bulk) Elements which are present in seawater and in the human body (in percent).

Element	Seawater ** (in percents)	Human body ** (in percents)	Human body (actual amount present in a man weighing 70 kilograms [154 pounds] in grams) *
Oxygen	33.0	65	45,500
Carbon	0.0014	18	12,600
Hydrogen	66.4	10	7,000
Nitrogen	- -	3	2,100
Chlorine	1.9	0.15	105
Sodium	1.05	0.15	105
Magnesium	0.135	0.05	35
Sulfur	0.0885	0.25	175
Calcium	0.04	1.5	1,050
Potassium	0.038	0.2	140
Bromine	0.0065	p +	p +
Strontium	0.0008	0.0002	0.140
Boron	0.00046	0.000014 or less	0.01 or less
Phosphorus	0.000007	1.0	700

* There are 454 grams in a pound, there are 1,000 milligrams (mg) in a gram, and there are 1,000 micrograms (μg) in a milligram. There are, therefore, 454,000,000 micrograms in one pound and 28,375 million micrograms in one ounce.

+ The letter 'p' indicates that the element was found to be present but that it hasn't been satisfactorily measured yet.

** Columns do not total 100 percent because we took data from diverse sources.

Table 1-2 adapted from Henry A. Schroeder, "Editorial, The Biological Trace Elements or Peripatetics Through the Periodic Table," **Journal of Chronic Diseases,** Vol. XVIII (1965), p. 218; Henry A. Schroeder, "Inorganic Metabolism," in Bernard L. Oser, ed. **Hawk's Physiological Chemistry,** 14th ed. (New York: McGraw-Hill, 1965), pp. 541, 561; E. J. Underwood, **Trace Elements in Human and Animal Nutrition,** 3d ed. (New York: 1971), p. 2; and Edward S. Deevey, Jr., "Mineral Cycles," in Board of Editors of **Scientific American,** eds. **The Biosphere** (San Francisco: W. H. Freeman, 1970), p. 85.

them. The human kidney is an example of an organ that is essential in detoxification. The kidneys filter poisonous substances (such as excessive copper) out of the blood and excrete them in the urine, but even more important, the kidneys keep or release the essential nutrient metal sodium (found in common table salt, sodium chloride), keeping the body's sodium supply constant within a narrow range of limits. Fresh river water doesn't contain much sodium, and the development of the kidney for homeostasis was an essential evolutionary development before mammals could migrate far from the edge of the sea.[33] As Henry Schroeder has said,

> Protoplasm was born in salinity and has never freed itself therefrom. All forms of animal life exist in a saline media, either free in water or carrying the media within impermeable or semi-impermeable skins. . . . When vertebrates first crawled on land, they met a set of conditions very different from the sea, not only in respect to water, but also to all of the bulk and trace elements contained in sea water. Concentrations and amounts available on land varied widely, and the organism had to reconstitute within its skin and tissues, from many and varied sources of food and water, the original media which cells require for health.[34]

Twenty-four elements are now known to be essential to animal nutrition. Table 1-4 shows the essential elements. The first four (hydrogen, carbon, nitrogen, and oxygen) make up 99.3 percent of the human body; the remaining twenty elements total only 0.7 percent of the body's atoms, yet we must stress that these minor elements are as essential as the bulk elements. Without them people (and other life forms) sicken and die. This is a fixed and firm rule of nature which cannot be changed: If the diet does not supply adequate quantities of all the essential elements, animals (including people) cannot remain healthy.

Carbon, hydrogen, oxygen, nitrogen, sulfur, and phosphorus make up the building blocks of living things, the amino acids, sugars, fatty acids, purines, pyrimidines, and nucleotides.

Seven elements (sodium, potassium, calcium, magnesium, chlorine, sulfur, and phosphorus) maintain the electrical neutrality of the body's fluids and cells.

Table 1-3. Minor (or Trace) Elements present in seawater and in the human body. Using percentages (as in Table 1-2), we are running up too big a string of zeroes for easy reading, so we must change the terms of measurement for showing the amounts of Minor Elements in humans and in seawater. Table 1-3 shows the amounts of these trace elements in parts per billion and parts per million (**ppb** and **ppm**). * (See also Table 1-2.)

Element	Seawater (ppb) *	Human body (ppm) *	Human body (actual amount present in a man weighing 70 kilograms [154 pounds] in milligrams) **
Fluorine	1300	p +	p +
Silicon	10 to 4,000	p	p +
Aluminum	3 to 2,400	1.4	100
Rubidium	120	17	1,200
Lithium	100	p +	p +
Iodine	50	0.43	30
Molybdenum	12 to 16	0.07 or less	5
Zinc	9 to 21	33	2,300
Barium	6	0.23	16
Lead	4 to 5	1.1	80
Selenium	4	p +	p +
Iron	3.4	57	4,100
Tin	3	0.43	30
Uranium	3.3	0.0003	0.02
Vanadium	2.4 to 7	0.3	20
Cesium	2	0.00014 or less	0.01 or less
Arsenic	1.6 to 5	1.4 or less	100 or less
Nickel	1.5 to 6	0.14 or less	10 or less
Titanium	1 to 9	0.21 or less	15 or less
Copper	1 to 25	1.4	100
Chromium	1 to 2.5	0.09 or less	6 or less
Manganese	0.7 to 1	0.3	20
Antimony	0.2	1.3 or less	90 or less
Cobalt	0.1	0.04 or less	3
Silver	0.15	0.01 or less	1 or less
Cadmium	0.03	0.43	30
Mercury	0.03	p +	p +
Gold	0.004	0.01 or less	1 or less
Thorium	0.001	?	?
Radium	0.3×10^{-10}	1.4×10^{-13}	10^{-10}

The so-called trace elements (fluorine, silicon, selenium, iodine, tin, zinc, copper, cobalt, iron, manganese, chromium, molybdenum, and vanadium) make up a third group of essential elements. Of these thirteen, ten are metals (selenium being considered a semimetal). As we have seen, these metallic elements participate in essential enzyme systems or in proteins with specific, vital functions, such as hemoglobin in human blood.

The iron in human blood is probably the best-known example of an essential metal. Iron in your blood helps form hemoglobin, which carries energy-giving oxygen throughout your body. In cases of dietary iron deficiency, your blood doesn't carry sufficient oxygen, you look pale as ashes because you're short of healthy hemoglobin, and you feel tired a lot of the time. (Iron deficiency anemia has been recognized as a specific human ailment since Roman times 2,000 years ago. The Romans drank rusty water for a cure.)[35]

It has only been during the past fifty years that medical science has come to understand the quantitative importance of trace elements in human health. (They are called trace elements because scientists 100 years ago could detect traces of them in the human body but couldn't measure their small quantities accurately.)

During the past two decades, and especially within the past eight to ten years, advances in laboratory equipment and techniques

---

* To illustrate the meaning of ppm and ppb, we should take the case of fluorine. There are 1,300 ppb of fluorine in seawater; this means that if we took a billion pounds of seawater and boiled it down and collected the residue, we would find 1,300 pounds of fluorine in the residue. The unit of measurement ppm works the same way. Take zinc in the human body, for example, where we find 33 ppm. This means that if we processed 1,000,000 pounds of human flesh and bones to extract the chemical elements, we would find 33 pounds of zinc.

+ The letter 'p' indicates that the element was found to be present, but that it hasn't been satisfactorily measured yet.

** There are 1,000 milligrams (mg) in 1 gram (g), and there are 454 grams in 1 pound.

Table 1-3: See sources under Table 1-2.

Table 1-4. Twenty-four essential elements in animal nutrition.

Element	Chemical Symbol	Atomic Number	Current Scientific Beliefs
Hydrogen	H	1	Essential to all life forms.
Carbon	C	6	Essential to all life forms.
Nitrogen	N	7	Essential to many plants and animals.
Oxygen	O	8	Essential to most life forms.
Fluorine	F	9	Essential for growth in rats; may be essential to human teeth and bones.
Sodium	Na	11	Essential to all life forms.
Magnesium	Mg	12	Essential to many life forms; found in chlorophyll.
Silicon	Si	14	Essential in chicks, perhaps in some plants too.
Phosphorus	P	15	Essential to all life forms.
Sulfur	S	16	Essential to all life forms.
Chlorine	Cl	17	Essential to all life forms.
Potassium	K	19	Essential to all life forms.
Calcium	Ca	20	Essential to humans.
Vanadium	V	23	Essential to some animals and plants.
Chromium	Cr	24	Essential in higher animals; related to action of insulin.
Manganese	Mn	25	Essential to most life forms.
Iron	Fe	26	Essential to most life forms.
Cobalt	Co	27	Essential to most life forms.
Copper	Cu	29	Essential to most life forms.
Zinc	Zn	30	Essential to most life forms.
Selenium	Se	34	Essential for function of liver in higher animals.
Molybdenum	Mo	42	Essential for higher animals and plants.
Tin	Sn	50	Essential in rats.
Iodine	I	53	Essential in thyroid hormones of animals.

Table 1-4 adapted from Earl Frieden, "The Chemical Elements of Life," **Scientific American**, Vol. CCXXVII (July, 1972), pp. 52-60.

have allowed great strides in the study of trace elements. As a writer in *Science* said recently:

> New research is continually expanding the number of metabolic systems known to be affected by trace elements and is revealing many hitherto unsuspected relationships between trace element concentrations and abnormal states of health. The time may thus be fast approaching when evaluation of trace element concentrations will play a fundamental role in the diagnosis of illness and when manipulation of those concentrations may play an even greater role in its prevention.[36]

As early as 1854, the French botanist Gaspard Chatin published his observations on the iodine content of soils, water, and food, stating his conclusion that the occurrence of goiter in humans must be associated with inadequate levels of environmental iodine. Goiter is a swelling of the thyroid gland at the base of the throat, the size of a goose egg in some cases. Goiter victims can be either dull and listless or nervous and edgy. Not all, but most goiter can be cured with dietary supplements of iodine.[37] Both sea salt and sea fish are excellent sources of essential iodine; as we have seen, life evolved from the sea, and our nutritional needs are still closely tied to our evolutionary beginnings. To make life possible inland, we now supplement most of our commercial table salt with iodine and prevent goiter in Midwesterners that way.

Many more serious diseases than goiter have been definitely linked to mineral deficiencies. A dietary copper deficiency prevents full development of blood vessels in rats. As a result these rats suffer from stroke—a weak blood vessel that bursts in the brain, causing death or severe disability.[38] Does a copper deficiency cause stroke in people?

According to a recent estimate, up to 23 million Americans now suffer from hypertension (high blood pressure). At least five metal nutrients have been linked to causation of hypertension. Likewise, a form of diabetes—a disease which frequently leads to lethal atherosclerosis (a kind of hardening of the arteries)—has been definitely linked to an imbalance of metal nutrients. Lung cancer, coronary heart disease, and emphysema have been found associated with abnormal balances of metals. Indeed, metals are intimately related to all our major killer diseases.[39]

Other specific human ailments known to be caused by dietary deficiency include beriberi, rickets, scurvy, night blindness, and kwashiorkor.[40] Anemia is a deficiency condition, which varies in seriousness from mildly enervating to lethal. Appendicitis and tooth decay are dietary ailments.[41]

Dr. George Cotzias joined other physicians and biochemists in the early 1960s in stating that many human degenerative diseases (the chronic diseases which plague old age) will soon come to be recognized as nutritional imbalances in metals. This prediction has steadily been coming true. In one generation a complex new frontier of medicine has opened up; the possibilities for human benefit from therapeutic nutrition have really just begun to be explored. What *has* been learned promises exciting new discoveries about the major killer diseases of the heart, kidney, and lung.[42] The badly neglected field of preventive medicine is commanding new attention. Paul Ehrlich (director of graduate studies in biology at Stanford) and his wife, Anne Ehrlich, recently wrote: "The importance of trace elements [in nutrition] cannot be over-emphasized. Because they are often required in such minute amounts, and perhaps because there seems to be considerable individual variation in requirements, there is some tendency to overlook them. Yet without them human life would cease."[43]

On the average, two new nutritionally essential trace metals have been discovered each decade for the past ninety years, with the bulk of the discoveries occurring in the past thirty-five years. During the two-year period 1970-1971, four new trace elements—two of them metals—became recognized as essential in animal and human nutrition.[44]

This, then, is the physical basis for understanding where we have come from. Now we will look at where we are and where we seem to be going.

> To maintain life in all its aspects, numerous chemical reactions must proceed in the body. These must go on at high speed and in great variety, every reaction meshing with all the others, for it is not upon any one reaction, but upon all together, that life's smooth workings must depend. Moreover, all the reactions must proceed under the mildest of environments: without high temperatures, strong chemicals or great pressures. The reactions must

be under strict yet flexible control, and they must be constantly adjusted to the changing characteristics of the environment and the changing needs of the body. The undue slowing down, or speeding up, of even one reaction out of the many thousands would more or less seriously disorganize the body.[45]

—ISAAC ASIMOV (1968)

2

Where We Are Now

The Microchemical Era of Environmental Health

PEOPLE began mobilizing metals and other chemical elements in a big way just 150 years ago, mining them out of the earth's crust, collecting them in central locations for processing, then distributing them widely (if unequally) according to prevailing beliefs about economic fair play, finally dispersing them into the biosphere (the living surface of the earth) after varied periods of delay.

For 2 million years people lived by hunting and gathering whatever the earth produced; just 10,000 years ago the practice of systematic agriculture developed; 4,000 years ago the first large urban civilizations took shape; soon people were getting into medium-scale mining, first lead, then copper. By 1100 B.C. iron was being mined. The nineteenth-century Industrial Revolution really began at least as early as the mid-sixteenth century.[1] During these early years important changes in habits of thought had to take place before the massive developments of the use of coal and iron and of steam power could be realized. People had to come to believe that quantitative growth by itself made human progress not only possible but inescapable. Once these beliefs became nearly universal, the success of industrial economics was assured, at least for the 150- to 200-year term. Now in the 1970s for the first time these old sustaining beliefs have encountered implacable challenges from informed men and women in the biological and physical sciences and in resource economics. Now it has become apparent to thousands (perhaps millions) of ordinary Americans that growth for its own sake is, as Edward Abbey once said, "the ideology of the cancer cell."[2]

By 1776 coal had become a major source of fuel in England, though wood remained prime fuel in the United States long past 1820 (when coal started coming into wide use over here). By 1870 world coal production had reached 250 million tons per year, and by 1895 the use of coal as fuel exceeded the world's use of wood as fuel for the first time.[3] In 1890 another fossil fuel—petroleum oil— began to be burned regularly.

Beds of coal and oil—the fossil fuels—were laid down in the deep earth hundreds of millions of years ago. Fossil fuels today release solar energy that was originally captured by photosynthetic plants aeons ago. The lush growth of primeval vegetation which eventually turned into beds of coal and oil originally grew following the same principles as modern plants grow, and so their cells contained a wide variety of minerals reflecting the local geologic environment during their ancient lifetimes. For this reason, both coal and oil contain approximately half of all of the ninety-two chemical elements known to occur naturally on planet Earth. Burning coal and oil (at between 2,000° and 3,000° Fahrenheit) releases many of these chemical elements directly into the atmosphere.[4] Lead, cadmium, mercury, nickel, arsenic, antimony, and vanadium, among others, become airborne this way. Even those elements which remain with the ashes will eventually be dispersed into the biosphere in other ways; a common practice today calls for covering coal and fuel-oil ashes with a layer of dirt. Such a dump site may leach chemicals into nearby surfaces or groundwater supplies for very long periods of time, probably exceeding centuries. According to recent estimates, people worldwide have mobilized and burned 154 billion tons of coal since the eighth century (and 34 billion tons of fuel oil since 1890).[5] Thus we can understand that even though trace minerals may appear in coal and oil at low levels of concentration, the total released from this source into the biosphere over the years has already been potentially significant. Fossil fuel combustion is a major source of contamination by mercury and other toxic metals in the surface waters of the industrialized nations of the world.

For example, if mercury is a constituent of coal and of petroleum at a concentration of 1 ppm* (which is probably low for an average), human combustion of coal and oil has released 377 million pounds of mercury into the biosphere since industrialization be-

---

* Ppm means "parts per million." Mercury in coal at the level of 1 ppm means there is one unit of mercury in each million units of coal.

gan. Depending on how it is distributed, this is a potentially significant quantity of toxic mercury.

In addition to major contamination from fossil fuel combustion, the industrial environment has now undergone thorough dousing with thousands of newly created synthetic organic chemicals, the best-known of them being pesticides. In the 1820s, just about the time people started burning large quantities of coal, European chemists began experimenting with the creation of synthetic organic chemicals. For many years chemists thought that only living organisms could manufacture the complex carbon-containing organic molecules upon which life is based. However, in 1828 a German chemist named Friedrich Wohler for the first time managed to create the right laboratory conditions for attaching carbon atoms to other atoms and thus manufactured the first synthetic organic molecule. He created synthetic urea, a nitrogen-based molecule found in urine. We should point out right away that just because a molecule is organic (carbon-containing), it doesn't mean it's necessarily good or healthy. Advertisers are stressing these days that their products are "organic"—but this just means carbon-containing and that's *all* it means.

During the 1850s a French chemist started systematic laboratory experiments to create other organic molecules, and he discovered that he could make chemical compounds that didn't exist in nature, brand-new compounds with a variety of marvelous qualities that could be exploited commercially.[6] By the 1880s Germany had developed a full-fledged chemical industry. The United States didn't take over leadership of the industry for another forty years.

Practical triumphs boosted the chemical industry tremendously throughout the late nineteenth century. The first synthetic organic chemicals were dyes. When chemists had looked for a source of carbon for experimental purposes, they had naturally hit upon coal as a good source. The first coal-tar dyes were a distinct improvement over earlier natural dyes; they were brighter, they lasted longer, and, after a relatively short time, they became much cheaper and more plentiful than their natural counterparts. (It was only later learned that many of them cause cancer.) In the 1860s, chemical laboratories produced a highly useful compound called dynamite. Among other things, this compound made it possible to blast a railroad bed across the United States. Practical triumphs like this assured the rapid growth of industrial chemistry (as well as industrial modes of warfare).[7]

By 1900, when General Electric set up the first corporate research laboratory, industrialists were becoming aware of the advantages to be gained from basic knowledge of the physical and chemical properties of matter. The first plastic—cellophane—had been developed before 1900; by 1909 Leo Baekeland had patented Bakelite and modern plastics were on their way. As a consequence of these initial successes, the chemical industry has grown strongly throughout the twentieth century. In the spring of 1971 the President's Council on Environmental Quality summarized the chemical situation this way: "About 2 million chemical compounds are known and several thousand new chemicals are discovered each year. Most new compounds are laboratory curiosities that will never be produced commercially. However, [300 to 500] of these new chemicals are introduced into commercial use annually. Of particular concern because of their rapidly increasing number and use are the metals, metallic compounds, and synthetic organic compounds."[8] The council went on to note that there are now more than 9,000 chemical compounds in commercial use in amounts over 1,000 pounds annually in the United States and that the total 1968 U.S. use came to nearly 120 billion pounds of chemicals (an increase of 161 percent over 1958 use). Table 2-1 shows the different classes of synthetic organic chemicals and how much of them the United States used in 1968 compared to 1967. The figures on organic chemicals do not include pesticides. In 1970 American chemical manufacturers produced 1.133 billion pounds of pesticides, and an additional 6 million pounds were imported from abroad; of this total amount, 730 million pounds were spread into the American environment and 409 million pounds were exported to be spread into distant environments.[9] The remainder went into reserve supplies to be dispersed into the environment later. This is an important point and one that frequently seems to be forgotten. Everything goes somewhere. After a delay period, sooner or later, everything we extract from the earth is spread into the biosphere. This is dictated by a principle of physics called the conservation of mass. Nothing can ever really "go away."

Many of the new synthetic organic chemicals are very persistent once they reach the biosphere. When we find natural organic molecules, we can be sure that nature contains enzymes which can unlock these organic materials and recycle them—break them down into their inorganic components, their original raw materials. This is why nature produces no unused wastes. Everything is ultimately

Table 2-1. U.S. production of synthetic organic chemicals, 1968 compared to 1967, not including medicinals or pesticides.

Chemical	1968 Production (in millions of pounds)	Percent increase over 1967
Intermediates	25,014	20.3
Colorants:		
Dyes	226	9.7
Pigments	54	1.9
Flavors and perfumes	117	4.5
Plastic products:		
Plastics and resins	16,360	18.6
Plasticizers	1,331	5.4
Rubber products:		
Processing chemicals	313	18.6
Elastomers	4,268	11.6
Surface active agents	3,739	7.5
Miscellaneous	67,525	13.1
Total	118,947	15.0

Table 2-1 adapted from Council on Environmental Quality, **Toxic Substances** (Washington, D.C.: U.S. Government Printing Office, 1971), p. 3. The following note appears below the original table: "Includes data on production measured at several successive stages in the manufacturing process and therefore reflects some duplication. Public disclosure is not permitted by the data-collecting agency when only one manufacturer produces a chemical. When production of an item was below 1000 pounds, or sales below $1000, a product is not included. Further, medicinals and pesticides are not included."

---

taken apart by enzyme action, and the basic inorganic components are reused.

The earth has not been encountering synthetic (human-created) organic molecules for very long; relatively speaking, they have appeared on the scene only in the most recent flash of history. For this reason, the natural systems of the earth have evolved no efficient means for breaking down many of these new organic compounds (such as DDT and toxaphene, to mention only two well-known synthetic organic compounds which are used as pesticides).

Because nature does not provide enzymes for breaking down many synthetic organic chemicals, they tend to persist in the environment. There, depending on their particular characteristics, they become dispersed and enter living systems. Since living creatures have never encountered these synthetic organic molecules before, they may or may not be able to cleanse their systems of any particular contaminant. When no homeostatic cleansing mechanism exists, continual low dosages of a contaminant can result in a buildup of toxic substances in a creature's body. People, for example, excrete DDT very slowly; that is why the average American today contains 12 ppm (parts per million) of DDT in his or her body. (This was no apparent cause for concern until the late 1960s and early 1970s, when it was shown that DDT is probably carcinogenic [cancer-causing].)[10]

Metals are even more troublesome than synthetic organic molecules because they can persist even longer in the biosphere. Metals don't degrade into harmless inorganic components; they are already basic inorganic materials, and they may or may not be harmless. At least fourteen metals are associated with serious environmental problems or human poisoning. Arsenic is probably the most famous. After they are dug out of the ground and dispersed into the environment, metals can and do remain active for very long periods of time. It is estimated, for example, that portions of the Great Lakes may be contaminated for as long as 5,000 years because of industrial dumping of the poisonous metal mercury.[11]

Like some synthetic organic compounds, some metals readily enter living things and move upward in the food chain or food web. The idea of the food chain is simple: the primary producers create food out of carbon dioxide, water, and essential elements using the energy of sunlight to drive the processes. These primary producers are eaten by some hungry creature, which will itself be eaten by a larger hungry creature. At the top of the food chain (three to five steps above the primary producers) are the top predators: large fish, large birds, bears, large cats, large dogs, and people.

When contaminants enter living systems, some of them tend to concentrate as they move upward through the food chain. This is sometimes called biological magnification. As a result, the top predators get a dose of poison which was contributed by thousands of primary producers and by other levels in the food chain. The top predators are therefore the ones that cen be expected to suffer the worst and earliest damage from pollution of the biosphere.

Since many people eat very high on the food chain, it is logical to look for changes in the human body attributable to industrialization. And when we look for such effects, we find them. What the rapid industrialization of the world has meant chemically is a large and increasing change in the trace element content of the average human over time. Through various means, we are unmistakably changing the trace element balance of the biosphere and of ourselves. To learn just how dramatic some of these changes in our bodies have already been, see Table 2-3. Notice, for example, that we have increased the amount of lead in our bodies by a factor of 170 and we have elevated our cadmium levels by a factor of 700. These are large, significant changes.

We are contaminating ourselves because we are contaminating the whole biosphere. For example, our mining, refining, using, and discarding of metals have now reached immense proportions on a global scale. Table 2-5 shows seventeen human mining activities which have now become so large that they dwarf natural, geologic erosion activities. When human activities begin to dwarf natural geologic activities in size, the potential for significant damage to the earth is very great. Look carefully at Table 2-5.

Worldwide mining activities are increasing at 5 percent per year. This means that total world mining production will double in the next fourteen years and double again in the fourteen years after that. Total world industrial production will double in the next ten years and double again in the ten years after that, growing at 7 percent per year as it is. World agricultural production is increasing at 3 percent per year; total world agricultural production is expected to double in the next twenty-three years and double again in twenty-three years after that.

If we try to arrive at some average total rate of growth of all human economic activities, we reach an answer of 5 to 6 percent per year. This means human impact on the global environment will be doubling every twelve to fourteen years now, multiplying at least fourfold by the end of this century. Such rates of growth are unparalleled and unprecedented in earth's history. Truly, nothing like us ever was. Humans just recently and quite suddenly have become an immense new force of geological magnitude for earth to reckon with.[12]

Table 2-2 shows world production for five toxic metals in 1968, and it shows U.S. consumption of the same metals for the same

Table 2-2. World production and U.S. consumption of five toxic metals (in thousands of metric tons) in 1968.

Chemicals	World production	U.S. consumption	Percent of world production represented by U.S. consumption
Cadmium	14.1	6.05	42.9
Chromium (as Cr_2O_3)	4730.	1200.	25.4
Lead	3000	1200	40.0
Mercury	8.81	2.60	29.5
Nickel	480.	144.	30.0
Total	8232.91	2552.65	31.0

Table 2-2 adapted from Study of Critical Environmental Problems, **Man's Impact on the Global Environment** (Cambridge, Mass.: MIT Press, 1970), p. 261.

year. It also shows the percentage of world production represented by U.S. consumption. In 1968 the United States used 31 percent of the entire world production of these five dangerous substances. (The U.S. population of 211 million people represents less than 6 percent of the present world population of 3.7 billion.) These and other metals reach the environment and the consumer from industrial manufacturing and processing sources, from smelters and refineries, from mine wastes, from the combustion of fossil fuels, from nuclear reactors, from municipal sewage systems, from the incineration and the land-fill dumping of solid wastes, from the combustion of gasoline and diesel fuel, from agriculture (cattle feedlots, pesticides, fertilizers, and irrigation and runoff waters), from food processing, and from numerous individual consumer products (prescription and nonprescription drugs, paint, clothing, and cosmetics, for example).

As we saw in our brief look at body chemistry, metals make excellent catalysts. As catalysts they initiate or change the speed of chemical reactions, in most cases accelerating reactions tremendously.

Table 2-3. Trace elements in the bodies of ancient primitive people compared to trace element contents of modern industrialized people, in parts per million (ppm).

Element	Primitive people (ppm)	Modern people (ppm)	Percent increase (+) or decrease (-)	Principal cause of difference
Copper	1.0	1.2	+ 20	Copper water pipes.
Bromine	1.0	2.9	+ 190	Bromides, fuel combustion.
Manganese	0.4	0.2	− 50	Refined foods.
Chromium	0.6	0.09	− 85	Refined sugars and grains.
Arsenic	0.05	0.1	+ 100	Feed additives, weed killers.
Vanadium	0.1	0.3	+ 200	Fuel combustion.
Lead	0.01	1.7	+ 16,900	Auto exhausts.
Cadmium	0.001	0.7	+ 69,900	Refined grains, water pipes.
Tellurium	0.001	0.4	+ 39,900	Metallurgy, tin cans.
Tin	0.001	0.2	+ 19,900	Tin cans.
Antimony	0.001	0.04	+ 3,900	Enamels, fuel combustion.
Mercury	0.001	0.19	+ 18,900	Fungicides, fuel combustion, fish.
Silver	0.001	0.03	+ 2,900	Eating utensils.

Table 2-3 adapted from Henry A. Schroeder, **The Trace Elements and Man, Some Positive and Negative Aspects** (Old Greenwich, Conn.: Devin-Adair Co., 1973), pp. 26, 27.

They work the same way in industrial applications as they do inside the human body. From its earliest days the synthetic chemical industry has used (and discarded) enormous tonnages of catalyst metals such as mercury and cadmium. Some fifty-two different metals are now mined for industrial use in the United States (or imported, in the cases where we lack deposits). Of these, fourteen have been identified by the Council on Environmental Quality as hazards to people and to the environment. The dangerous fourteen are: arsenic, barium, beryllium, cadmium, chromium, copper, lead, manganese, mercury, nickel, selenium, silver, vanadium, and zinc. Table 2-4 shows the U.S. consumption of the fourteen toxic metals in 1948 and 1968. The increases are dramatic.

Table 2-4. Estimated U.S. consumption of poisonous metals, 1948 and 1968, showing percentage of change over the period.

METAL	Total estimated consumption (in tons) 1948	1968	Percentage increase, 1948 to 1968
Arsenic	24,000	25,000	4
Barium	894,309	1,590,000	78
Beryllium	1,438	8,719	507
Cadmium	3,909	6,664	70
Chromium	875,033	1,316,000	50
Copper	1,214,000	1,576,000	30
Lead	1,133,895	1,328,790	17
Manganese	1,538,398	2,228,412	45
Mercury	1,758	2,866	63
Nickel	93,558	159,306	70
Selenium	419	762	82
Silver	3,611	4,983	38
Vanadium	Not available	5,495	- -
Zinc	1,200,000	1,728,400	44

Table 2-4 adapted from: Council on Environmental Quality, **Toxic Substances** (Washington, D.C.: U.S. Government Printing Office, April, 1971), p. 2.

Five metals have been ranked by Dr. Vaun Newill of the federal Environmental Protection Agency (EPA) as particularly dangerous air pollution health hazards. Dr. Newill ranks the five in this order of importance: cadmium, lead, nickel, beryllium, and antimony.[13] To this list, Dr. Henry Schroeder adds tellurium, if not necessarily as an air pollution hazard, still as a general danger to people.[14]

Twenty-two different metals and countless organic compounds now can be measured in the air of all our cities. As Dr. Schroeder has calculated:

In every cubic kilometer of air over North American cities—which is not very much air as air goes, the cube being about 1,000 yards on a side—there are 6 to 9 pounds of aluminum, 6 to 33 pounds of iron, 2 to 15 pounds of magnesium, 2 to 18 pounds of sulfur, 9 to 13 pounds of

Table 2-5. The size of human mining activities compared to size of global erosion. Generally speaking, elements with highest A/B ratios are greatest potential polluters of water.

Chemical symbol	Name of element	A Amount mined (kg per year)	B Amount added to oceans by rivers (kg per year)	Ratio of A to B
Ag	silver	6.5×10^6	4.8×10^6	1.4 / 1
Al	aluminum	1.1×10^{10}	8.9×10^9	1.2 / 1
Ca	calcium	5×10^{11} ?	5.7×10^{11}	1 / 1 ?
Cr	chromium	2×10^9	6.7×10^6	300 / 1
Cu	copper	4×10^9	3.7×10^8	11 / 1
Fe	iron	2.1×10^{11}	2.5×10^{10}	8 / 1
Hg	mercury	1×10^7	3×10^6	3 / 1
Mn	manganese	6×10^9	4.4×10^8	14 / 1
Mo	molybdenum	3×10^7	1.3×10^7	2.3 / 1
Ni	nickel	3×10^8	3×10^8	1 / 1
P	phosphorus	1.4×10^9	1.8×10^8	8 / 1
Pb	lead	2.2×10^9	1.8×10^8	12 / 1
Rb	rubidium	9×10^7 ?	5.5×10^7	1.6 / 1 ?
Sb	antimony	4.5×10^7	1.3×10^6	35 / 1
Sn	tin	1.7×10^8	1.5×10^6	110 / 1
Ti	titanium	1×10^9	3.2×10^8	3 / 1
Zn	zinc	3×10^9	3.7×10^8	8 / 1

Table 2-5 adapted from: H. J. M. Bowen, **Trace Elements in Biochemistry** (New York: Academic Press, 1966), pp. 163, 164.

---

silicon, 0.4 to 4 pounds of zinc, up to 2 pounds of copper, up to 2 pounds of titanium, 1 to 7 pounds of lead, and smaller amounts of arsenic, beryllium, fluorine, manganese, nickel, antimony, tin and vanadium.[15]

In addition to appearing suspended in the air, metals and synthetic organic chemicals can be measured as contaminants throughout the nation's surface and underground water supplies.[16] These trace pollutants are not removed by ordinary municipal water treatment. "Because trace organics are not removed from surface water by ordinary treatment practices and subsurface waters are not usually treated in any manner, a serious health threat exists to the consumer by the presence of these materials," say three investigators, who, to their surprise, found synthetic organic chemicals polluting deep groundwaters in Missouri.[17] In April 1975, the EPA announced finding synthetic organic chemicals polluting the drinking water supplies of all 79 American cities which were surveyed. Among other chemicals found in drinking water supplies were chloroform and carbon tetrachloride, both suspected carcinogens.[18] When the U.S. Public Health Service checked tap water at 2,595 places in the United States in 1969, it found that 36 percent of the samples exceeded drinking water standards for one or more contaminants. Arsenic, iron, lead, manganese, and cadmium appeared in some of the samples at concentrations ten times higher than the Public Health Service safety limit.[19]

There are at least 496 different organic chemicals known or suspected to be contaminating large portions of the nation's water supplies now. Thirty-three of these chemicals, which have actually been found as pollutants in one water supply or another, have been tested to see if they are carcinogenic. Of the thirty-three, fifteen were judged to be cancer-causing agents. Of another eighty-seven organic chemicals, tested because it was thought they were probably contaminating water supplies extensively in the United States, seventeen of the eighty-seven (20 percent) were found to cause cancer.[20]

There are now about 1 million different consumer products manufactured in the United States. Most of the major categories of consumer items now have chemicals put on them or in them; most of our clothing, most of our housing and building materials, and now even most of our foods are intentionally treated with chemicals. More than 3,000 different food additives are now intentionally put into the American diet. Fifteen years ago we used a total of 400 million pounds of food additives; in 1970 we used 800 million pounds. In 1970 for the first time Americans ate more processed foods than fresh foods.[21] Since 1940 the use of food coloring (frequently coal-derived dyes) has increased twentyfold.[22] The trend is clear.

The national food supply is also being unintentionally contaminated. In 1965 the *Report of the Environmental Pollution Panel of the President's Science Advisory Committee* said:

> Soils are being polluted with a variety of substances, both inorganic and organic. Although problems arising from contamination have been recognized for some time, neither the severity nor the magnitude of the broad issue of soil pollution has been widely appreciated. . . .
>
> It is clear that ever-increasing amounts of pollutants have been entering the soil in recent years. Radioactive fallout, organic pesticides, radionuclides in fertilizers, heavy metals, salts from irrigation water, industrial and household wastes, salts used on roads for the purpose of de-icing, lead from automobile fuel combustion, pathogenic microorganisms, growth-regulating chemicals and mulching materials represent some of the more readily apparent soil contaminants; others may be recognized.[23]

The 1965 *Report* went on to say, "Focussing [sic] attention upon soil pollution is unpopular in many circles, but the issue has now become a matter of national concern. The land resources available to us are too precious and too limited to permit delay in undertaking a major effort directed at the prevention, abatement, and control of the contamination of the soils of the United States."[24]

A medical warning was sounded 50 years ago, just about 100 years after the chemicalization of the environment got under way. A physician named William Salant, writing in the *Journal of Industrial Hygiene* in June 1920, noted that metals were now found everywhere in America, especially in the food supply. He surveyed current information about metals, revealing enormous lack of knowledge in the professional literature. He surveyed the occupational health situation—finding metals a definite problem there—and speculated about the long-term health effects of subjecting the American people to chronic, low-level exposures to metals (and, we add, to other chemicals). "The effect of heavy metals on health is a matter of prime importance," wrote Dr. Salant, mentioning lead, mercury, cadmium, vanadium, manganese, chromium and nickel.[25]

All informed people now recognize somewhat too late that Dr. Salant was right. As *Business Week* said on May 11, 1974:

> After three decades of unprecedented and virtually un-

regulated proliferation of new chemical products, American industry now confronts a frightening fact. . . . Long-term exposure to an unknown number of chemicals can produce irreparable damage to the organs of employees who work with them—and chemicals are used in every nook and cranny of U.S. industry, not just the $70-billion, 1-million-worker chemical industry.

All of these chemicals eventually find their way out of the workplace, as we shall see.

The chemicalization of the environment proceeded slowly but steadily throughout this century until World War II, when the *rate* of chemicalization stepped up. Our use of fossil fuels is doubling every fourteen to eighteen years now,[26] and our production of synthetic chemicals is doubling every seven years or less. After decades of complacent progress, suddenly it has become apparent that the microchemical era of environmental health rushed upon us from the horizon before most of us even saw it looming. Now we are just beginning to understand the consequences. "It would seem not unlikely that we are approaching a crisis that is comparable to the one that occurred when free oxygen began to accumulate in the atmosphere," said Yale biologist G. Evelyn Hutchinson in 1970.[27] As we saw in Chapter 1, the oxygen crisis, when it occurred perhaps 2 billion years ago, extinguished the vast majority of all existing life forms on earth.

The greatest single change in man's surroundings during the past several decades has been the ever increasing chemicalization of his environment. The human habitat is becoming more and more menacing. It is estimated that there is now six times as much pollution in our rivers and lakes as there was 60 years ago and the amount is still increasing. Every year, more than 500 new chemicals are introduced into industry, along with countless operational innovations. Professor J. McKee at the California Institute of Technology, in his recent address to the National Research Council, indicated that "We have entered the micro-chemical era of environmental health."[28]
—D. W. Ryckman and others (1968)

3

Lethal Potentialities

Cadmium

FOR more than 2 billion years life developed in the oceans. During all that time there was very little poisonous cadmium around. See Table 1-3.[1] But now contamination by this highly toxic industrial metal can be detected almost everywhere. Today it is known that cadmium enters our bodies (through air, water and food) in sizable quantities, averaging between 70 and 200 micrograms per person per day in the U.S.[2] And there is strong evidence that cadmium is an important contributor to heart disease, the nation's No. 1 killer. Dr. Henry Schroeder says flatly that cadmium is "a present and real hazard" to the American people.[3]

Geologically, cadmium is very close to zinc; zinc and cadmium always appear together in nature. Geologically, the two metals are inseparable (though modern chemists can separate them). This is an important point, as we shall see. In seawater, for every atom of cadmium there are about 100 atoms of zinc.[4]

Unlike cadmium, zinc has always been an essential element.[5] As we saw in a previous chapter, when the human body developed from earlier life forms, organic molecules had already taken up the habit of building proteins using zinc as a catalyst. The zinc steps into the necessary reactions, helps string the amino acids together into proteins of the right combination and configuration, then steps back out of the reaction. The empirical formula of the resulting protein does not show zinc as a constituent part but zinc was essential in the protein's creation. Therefore zinc has always been an essential element in the human body. As we shall see, zinc is essential in almost every biochemical system, and a zinc deficiency has recently been measured and described among a group of affluent Caucasian children in Denver, Colorado, which is to say in average well-to-do American children.[6]

Animal bodies must always take in (from plants and water) enough zinc to keep cells healthy. To balance out the highs and lows of dietary zinc intake, the human body developed a homeostatic mechanism (which relies on the same feedback principle as your home thermostat) to regulate absorption and excretion of this valuable metal. The body of a 70-kilogram (154-pound) man will ordinarily contain 2.3 grams (0.08 ounces) of zinc. We have noted that zinc and cadmium always appear together in nature; therefore, the human body has always encountered a small amount of cadmium when it has ingested its essential zinc.

Unfortunately the body has, historically, encountered so little cadmium that no homeostatic mechanism has developed to regulate the body's cadmium content. The human body has never developed any efficient means for excreting cadmium because, on the evolutionary time scale, the body has never needed to excrete much cadmium.[7]

When the human body encounters cadmium, the metal is rapidly transferred from the stomach or bloodstream into the kidneys and liver (mostly the kidneys). There the cadmium is stored tightly away and it doesn't leave the body for a very long time. The half-life of cadmium in humans ranges from ten years to twenty-five years;[8] this means, if you eat a certain amount of cadmium today, you will still have half of that amount left in your body ten to twenty-five years from now; in another ten to twenty-five years you will get rid of half of the remaining half, and so on. A half-life of ten to twenty-five years is relatively long, and it assures a buildup of cadmium in the bodies of industrialized peoples unless very strict control measures are enforced. Every time we take in a little cadmium, the body just adds it to the growing cadmium reserve in the liver and kidneys.

The kidneys are two fist-sized organs located in your back, just below the lowest ribs. These indispensable organs filter the blood constantly, removing waste materials and poisons. The kidneys work twenty-four hours a day, constantly keeping the blood clean and constantly regulating the levels of many essential elements (such as sodium). If you take in a large quantity of sodium (the met-

al found in sodium chloride, table salt), your kidneys rapidly remove precisely enough sodium to keep your blood level within tol-
erable limits.

Human beings are ordinarily born without any cadmium in
them, or with not more than about 1 microgram (μg) total. However, the average breast-fed infant in America today begins ingest-
ing cadmium immediately (at the average rate of 3.65 μg per kilogram* of body weight each day).[9] As we age, cadmium accumulates
in the kidneys. We must stress that the filtering actions which the
kidneys are constantly performing are an exquisitely delicate and
selective series of operations.[10] Separating waste materials and poisons from the bloodstream requires actually disassembling portions
of the blood and reassembling it through an intricate network of
140 miles of tubes and filters. Every 24 hours this vast network han-
dles 200 quarts of blood, removing 2 quarts of useless or poisonous
materials which are excreted from the body as urine.

Up to a point, cadmium apparently doesn't interfere with the
kidneys. Attaching itself to a protein called metallothionein, the
cadmium remains relatively inert.[11] Binding to metallothionein
may be the body's homeostatic mechanism, its attempt to protect itself against the toxicity of cadmium. However, as cadmium builds
up more and more in the kidneys, eventually damage can begin.
The homeostatic mechanism is overwhelmed. *When* damage begins
apparently will depend on *how much zinc is present in the kidney.* Here
is a clear example of the balance of elements at work. Cadmium has
the ability to displace essential zinc in the kidney. Through careful
measurements, Dr. Henry Schroeder and his coworkers have
found that when cadmium builds up and reaches a certain ratio to
zinc in kidneys, then damage in animals (and humans) can be expected.[12] The kidney damage can show up as high blood pressure.
High blood pressure is a contributory cause to stroke and to many
forms of heart disease, and heart disease kills nearly 60 percent of
all the Americans who die each year. So the cadmium–zinc balance
in the kidneys is an important public health matter. Obviously, the
ratio of cadmium to zinc can be changed for the worse either by
dietary zinc deficiencies or by cadmium contamination of air, food,
and water. The *balance* of elements is what's most important.

If we visualize the total time humans have been on earth (2 mil-

---

*A kilogram is about 2.2 pounds.

lion years) as a straight line 200 feet long (two-thirds of a football field), then the total time since the beginning of large-scale combustion of coal (150 years or so) can be visualized as a line measuring 0.2 inches in length, or about a quarter of an inch. This is the reason our bodies are now experiencing difficulty adjusting to the increase in cadmium levels in our environment: on a geological, evolutionary time scale, our environment has been loading up with cadmium for only the briefest instant. Our genetic mechanisms for adjusting to environmental change require much longer periods of time to adapt to new conditions.

It was just 120 years ago that German businessmen began selling cadmium in a powdered silver polish. By the 1870s German chemists had discovered that cadmium made good pigments for red, yellow, and orange paint, and by 1900 something like 100,000 pounds a year were going into annual use worldwide. U.S. production began in 1907, but the German chemical industry led world consumption until 1919 or so, when the United States took the lead it has maintained ever since.[13] In 1968 Americans "consumed" 6,664 tons of cadmium. (See Table 1-2 in Chapter 1.) The word "consumed" is misleading, however, because, as the President's Council on Environmental Quality has pointed out, "The basic principles of physics dictate that all materials used in the productive and consumptive activities of the economy must eventually be returned to the environment."[14] Instead of saying that Americans "consumed" more than 13 million pounds of cadmium in 1968, we should speak of "mobilizing" all that cadmium and slowly, after varied periods of delay, dispersing it into the general environment.

Long after we started mobilizing large quantities of cadmium and dispersing it into the biosphere, we began to understand some of cadmium's poisonous qualities. It was during the 1920s that a body of medical knowledge about cadmium began to develop. Long ago the Romans had recognized it as toxic. In modern times, cadmium killed its first reported industrial/consumer victim in 1858 (a woman had been polishing silver with a cadmium compound in closed quarters; she developed a lung disease like emphysema and died).[15] However, it was not until World War I that we learned about many of cadmium's dangerous industrial properties. When cadmium is heated above 321° Centigrade (610° Fahrenheit) it melts; if it is heated much beyond that, it ignites, actually catches fire, and gives off a brown or brownish-yellow fume which has been tested on dogs and found to be twice as poisonous as phosgene gas. (Phosgene is the well-known poison gas developed for use in World War I.)[16]

For a time military authorities considered using cadmium fume as an agent of chemical-biological warfare (CBW).[17] From a military viewpoint, some of the behavior of cadmium fume suits it ideally for CBW use. At first the fume isn't particularly irritating to the nose or throat; it does not even have a pronounced odor. Yet at varying doses it attacks many different parts of the human body, destroying the inner surfaces of the lungs, producing the effect of gangrene in the testicles, obliterating nerve tissue in the brain and spinal cord, and finally causing large parts of the kidneys to slough off and pass from the body with the urine.[18] Cadmium is an enzyme poison par excellence. In addition, cadmium has some delayed effects which might only show up in a target population long after the battle had ended. For one thing, in high doses cadmium can cause cancer. For another, it has teratogenic effects upon pregnant women (literally "monster-producing" effects, causing birth defects). For the 100-year effect on the enemy, cadmium may be mutagenic—causing gene mutations which can show up two to five generations later in the form of spontaneous cancer, or weak eyes, or a bad back, or kidneys that don't function up to par—or any of a thousand other diverse maladies.[19]

However, cadmium was never developed as an agent of CBW (although another toxic metal, arsenic, was), perhaps because its initial effects are not immediately felt; there is usually a delay of several hours before the onset of symptoms in acute (immediate, short-term high-dosage) poisoning. In cases of chronic (delayed, long-term low-dosage) poisoning the effects usually take at least two years to come on.[20] The effects of chronic, low-level cadmium poisoning, when they do appear, may be varied. A lung disease like emphysema is a common effect; this progressive shortening of breath starts like asthma or chronic bronchitis and leads, within a few months to a few years, to severe disability. People with advanced emphysema cannot walk more than about ten yards without feeling exhausted. Cadmium lung disease produces similar debilitating effects.

In addition, chronic, low-level cadmium exposure (as among groups of industrial workers, many of whom frequently don't know and aren't told they're working with cadmium alloys or cadmium-plated materials) can lead to anemia and kidney disease, including kidney stones, then to complications of the cardiovascular system (heart disease or high blood pressure) and finally to death.[21] In the workplace, and among welders and crafts people who use a lot of solder, cadmium is an extreme hazard.

Despite its failure to develop as an agent of CBW, cadmium nevertheless emerged from World War I as a major industrial metal, mostly for plating iron and steel. During the war years cadmium found widespread use as a bright, silvery rust-preventive coating on ferrous (iron-containing) metals. U.S. production rose steadily from 14,000 'pounds in 1907 to 207,000 pounds in 1917. After the war cadmium remained cheap and useful until, by 1962, U.S. production had reached 11 million pounds per year. Between 1907 and 1963 American industries and consumers released an estimated 250 million pounds of cadmium into the environment without bothering to keep track of it, not including the unknown tonnages released by burning fossil fuels. Here is a brief discussion of the ways in which cadmium reaches the environment (and, after a delay, human beings):[22]

Mining: Cadmium does not appear in nature as a distinct substance with its own ore; instead, we obtain cadmium by refining zinc, lead, and copper ores. In other words cadmium is a by-product of the mining of those other three metals. Our biggest single source of cadmium (besides imports from other countries) is zinc ore. Zinc, lead, and copper mines produce dust containing cadmium which eventually reaches the atmosphere, then the soil and water but the total amount emitted into the atmosphere from this source appears to be small. After mining operations have been abandoned, should the mines fill with water (as mines frequently do) a serious mine-drainage water pollution problem can develop. Mine wastes in Missouri contain up to 1000 ppm cadmium.[23] Piles of mine tailings (leftover crushed rock) and previously strip-mined areas (where vegetation is usually very reluctant to grow) can present a continuing leakage of cadmium into the general environment. Mining associated with cadmium is now carried out in twenty-six states. A total of 5,304,000 pounds of cadmium came out of domestic mines during 1968 (though we should note that this represented only 38 percent of all ore processed in the United States in 1968, the other 62 percent having been imported).

Ore Processing: After ore has been removed from the earth, it is usually heated to drive off the metals; these are collected in a concentrated form and sold. Because we use more cadmium than we produce, we import cadmium-containing zinc and lead ores from other countries (Canada, Mexico, Peru, Morocco, Honduras, Bolivia, South Africa). We also import flue dusts which have been collected from the smokestacks of foreign lead and zinc smelters (which gives us an indication of how much cadmium escapes from uncontrolled smelters).

Importing these ores and flue dusts (containing 6,208,000 pounds of cadmium metal in 1968) presumably doesn't pollute the atmosphere or water or soil during transport, though one would expect loading and unloading to be a dusty (therefore hazardous) job. However, *processing* these ores contributes much cadmium to the general environment. An estimated 2,100,000 pounds of cadmium reached the U.S. atmosphere from ore processing in 1968. How much reached the water from this source remains unknown, though it is certain that *some* cadmium went this way. Ore-processing plants may dump waste water into a river or lake or the ocean (or they may divert it into a holding pond to delay its return to the environment). When cadmium enters a municipal sewage system (from, say, an electroplating operation), it is mostly retained in the sewage sludge. From there it may be spread on city parks for fertilizer or it may be sold to a farmer for fertilizer. When it is used on parks, worms and other organisms in the soil will pick up some of the cadmium, birds will eat the worms, and cadmium does in this way enter the wildlife food web. When sewage sludge is used to fertilize crops, the crops will pick up some of the cadmium, and people (or cattle, then people) will eat it. Ore processing is the single largest source of cadmium to the atmosphere (and thus to the general environment).

From all sources (domestic mining plus import of flue dusts and ores) the United States produced 10,225,000 pounds of new cadmium metal in 1968. In addition, the government released 808,000 pounds from its cadmium stockpile, private industry released 372,000 pounds from its stockpile, and we imported 1,927,000 pounds of new metal (mostly from Australia, Canada, Mexico, Japan, and Peru). In addition, we recycled 426,000 pounds of old cadmium (mostly from alloys). We exported 530,000 pounds of new cadmium metal to foreign markets, leaving us a grand total of 13,328,000 pounds to put into the manufacturing systems of our nation (and from there, ultimately, into our general environment).

NASA physicist Charles Hyder estimates that combustion of coal contributes perhaps 2 million pounds of cadmium to the environment each year, and, he says, "I would expect oil burning to raise that to more than 20 million pounds annually."[24]

THE CONSUMERS OF CADMIUM

Plating iron and steel with bright, lustrous cadmium not only protects the iron and steel against rust but also improves its appearance, in the opinion of many. Hundreds of thousands of steel consumer items are cadmium-plated: screws and bolts, miscellaneous hardware items (cotter pins, wire, door hinges, you name it), thousands of automobile and aircraft parts, electrical and electronic apparatus (switches, chassis for TVs and hi-fi sound systems), and diverse household appliances. Cadmium makes an especially good protective coating in harsh environments, near the seashore or in a tropical atmosphere where rust and corrosion rates are high.

During 1968, 45 percent (or 6 million pounds) of all our cadmium went into the electroplating of industrial equipment and consumer items. These items will remain in use for up to twenty years (in rare cases, longer) and will then be discarded into municipal dumps or incinerators; air pollution will result from the incineration and water pollution from the landfill dumping.[25] Some cadmium-plated steel items will make their way back to a scrap yard and be recycled to a steel mill; from here, most of the cadmium will reach the atmosphere. In 1968 the American steel industry purchased 38,500,000 tons of scrap steel which contained an estimated 2,040,000 pounds of cadmium (from cadmium-plated and zinc-galvanized steel among the scrap). Of this cadmium, an estimated, 2,000,000 pounds escaped into the air when the scrap steel was melted down. This is the nation's third-biggest source of airborne cadmium, after fossil fuel combustion and ore-processing plants (smelters).

Another important use of cadmium is pigments, colors for paints and plastics. Cadmium pigments were first developed in Germany slightly over 100 years ago. However, this did not become a major use of cadmium until 1930. Early in the 1930s, chemists discovered that a combination of cadmium, sulfur, selenium, and barium made possible a range of bright pigments, from lemon yellow through deep yellow, orange and red and on into deep maroon. The color depended on the mixture of elements. In addition to the bright, rich colors, these pigments have turned out to be extremely long-lasting. Neither time nor heat destroys their depth or brightness. These pigments quickly found wide use in durable enamels and other finishes, in plastics (especially the popular polyvinyl chloride, or PVC), and in coated textiles (so-called "oil cloth," for example, or "wet look" clothing), rubber, colored glass, printing inks, baking enamels, ceramic glazes, and artists' colors. *Many household dishes, intended for the dinner table, contain dangerous cadmium glazes and colors and decals.* The federal protective agencies have thrown up their hands in dismay, saying this dinner-plate pollution is now so widespread that they can't solve it.[26] Don't buy plates or bowls which are bright orange, yellow or red; acidic foods coming in contact with these finishes may leach cadmium. Caveat emptor.

Cadmium is added directly to many plastics—especially to polyvinyl chloride (PVC)—to make them stronger and make them last longer in service to the consumer. In this use cadmium is not a pigment but a stabilizer. Plastic bottles, plastic tubing, plastic pipe, plastic garden hose, plastic rope, plastic lawn furniture webbing, phonograph, hi-fi and stereo records, shower curtains, clear vinyl packaging (except food wrappers, where use of cadmium stabilizers in the plastic is forbidden by U.S.D.A.), plastic tabletops, wall covering, house siding, automobile upholstery, toys—these are just a few of the plastic products that frequently contain cadmium (most frequently the ones colored red, orange, or yellow, though PVC water pipe comes in black and white and can contain much cadmium). All together, U.S. industry employed 2,800,000 pounds of cadmium in pigments in 1968 (or 21 percent of all U.S. cadmium used that year). An additional 2 million pounds went into plastics as stabilizers.

•Cadmium mixed with silver makes an exceptionally strong soldering material; such alloys vary from 10 to 40 percent cadmium content. Brazing with a cadmium-silver combination makes it possible to join ferrous and nonferrous metals in strong leakproof joints which resist corrosion—an important industrial use. Such brazing strips usually contain 15 to 25 percent cadmium. It is likely that cadmium solders and brazing materials are used to close up containers and equipment in the food-processing industry, which explains to some degree why cadmium can be measured in soft drinks and institutional diets and, indeed, in everyone's diet. (See Table 3–1.)

•Batteries. Nickel-cadmium dry cell batteries have every advantage over their main competitor (lead-acid batteries) except price. Nickel-cadmium batteries last longer, don't usually burst at the seams, can be recharged hundreds of times (and quite quickly), and are not affected much by temperature. They also deliver maximum current with minimum voltage drop. Most of these dry cell batteries today end up in landfill dumps, where they are ultimately liable to pollute water.

•Electroplating, pigments, plastics, alloys, and batteries consumed 90 percent of U.S. cadmium in 1968. The remaining 10 percent was used in small quantities by hundreds of manufacturers of fungicides (more than 25,000 pounds of cadmium metal was spread onto golf courses during 1968). Cadmium control rods, thrust into the core of a nuclear reactor, absorb neutrons and slow the reaction. Cadmium makes good phosphorescents for TV tubes and X-ray screens, and it can lend an iridescent effect to glazes on pottery and porcelain (look for it on brown coffee mugs). Other miscellaneous uses include photographic chemicals, the curing of rubber (though not for automobile tires), lithography, process engraving, and glass manufacturing.

Cadmium reaches the environment and the consumer from other ways, several of which may be very important and a few of which appear to be relatively minor. Even the minor sources, however, can teach an important lesson: Metals turn up in unexpected places. Take, for example, automobile tires. No cadmium is intentionally added to automobile tires. Yet tires are made using a zinc catalyst which may itself be contaminated with cadmium. New rubber tires have been measured, and they contain from 20 to 90 parts per million (ppm) of cadmium. This means that, if you burned 1 million pounds of rubber tires and collected the smoke and ashes, you would find 20 to 90 pounds of cadmium. Since tires wear down as we drive on them, and since the resulting dust enters the atmosphere, auto tire wear contributes an estimated 11,400 pounds of cadmium to the general U.S. environment each year. (This does not include cadmium put into the atmosphere by burning tires.)

Altogether automobiles put an appreciable amount of cadmium into our atmosphere each year, the total amount remains uncalculated; it hasn't even been officially *estimated* because it's not supposed to be happening. Automotive lubricating oil is thought to contain cadmium at the level of 0.48 ppm. From this source our cars burn (and emit from their exhausts) an estimated annual 1,820

Table 3-1. Cadmium levels measured in food, beverages, soil, vegetation and selected commercial products.

Material	Concentration in parts per million
BEVERAGES, DAIRY PRODUCTS	
Coffee, ground	0.32
Coffee, instant	0.06
Coffee, instant, dried	2.27
Coffee	0.35
Tea, infusion	0.01
Tea, Japanese, green leaves	2.50
Tea	0.32
Milk, homogenized (3 samples)	0.10–0.14
Emmenthaler cheese	1.48
Bourbon whiskey	0.10
FRUITS, VEGETABLES, ROOTS	
Beans, string, fresh	0.01
Potatoes	0.03
Potatoes, German (3 samples)	0.18–0.20
Carrots	0.30
Spinach, fresh	0.45
Tomatoes	0.03
Tomatoes, Holland	0.25
MEAT	
Beef liver	0.28
Beef kidney, German (3 samples)	12.00
SEAFOOD	
Oysters, frozen	3.14
Oysters, fresh	3.66
Anchovies, canned	5.39
Tuna, canned in oil	0.20
Saltwater fish, canned (4 samples)	0.15 – 0.20

Material	Concentration in parts per million
CEREALS, GRAIN PRODUCTS	
Wheat, whole	0.25
Bread, whole wheat	0.15
Bread, white	0.22
Corn, hybrid	0.12
Rice, brown	0.04
Rice, Japanese, polished	0.06
Cereals and cereal products, German (13 samples)	0.16
OILS AND FATS	
Olive oil, Spanish	1.22
Cod liver oil	1.71
Castor oil	1.26
Cooking fat, German (17 samples)	0.14
VEGETATION (from wild forest)	
Leaves	
Pine	0.05
Oak	0.09
Beech	0.04
Fungi	
Amanita muscaria	0.71
Phallus impudicus	0.42
SOIL	
Soil, under maples, wild forest	1.20
Soil, under oaks, wild forest	2.45
Soil, fertilized (Virgin Islands Agricultural Station)	3.38

(Continued, on next page)

pounds of cadmium. Automotive gasoline, on the other hand, is not supposed to contain cadmium at all. Yet Ralph Nader's Center for Study of Responsive Law took twenty-one brands of gasoline to an independent testing laboratory, and results revealed cadmium in four brands (19 percent). The Nader-sponsored study showed 35 parts per billion (ppb) of cadmium in Enco and Chevron gasolines, 30 ppb in Shell, and 10 ppb in Texaco.[27] Since Texaco and Shell have the nation's first and second largest gasoline sales, these findings are noteworthy. If this cadmium is being deliberately added to gasoline (in violation of the law) without registering cadmium as an additive, the practice can at least theoretically be controlled.

If, however, the cadmium is entering the gasoline as a contaminant, the control problem becomes more difficult. Many pipes, tank trucks, and storage tanks may be plated or welded with a cadmium-containing alloy; rubber or plastic hoses may also contain cadmium. If 20 percent of the nation's gasoline emits cadmium into the air when it's burned, we have a problem—from this one source alone—that has reached "the questionable hazard stage," according to Dr. Henry Schroeder.[28]

The toxic actions of cadmium are poorly understood. It *is* known that cadmium inhibits enzymes and that this is the chief way in which it poisons people but saying an enzyme has been inhibited and knowing precisely how it happened at the microchemical cellular level are two entirely different things. It *is* known that cadmium can attack virtually every system of the human body; that is why

Material	Concentration in parts per million	Material	Concentration in parts per million
SOIL (continued)		FERTILIZERS, ROCKS	
Soil, tilled, unfertilized (Virgin Islands Agricultural Station)	0.80	Superphosphate fertilizer, U.S.	8.97
Field, unfertilized (Virgin Islands Agricultural Station)	0.15	Ground phosphate rock, Tennessee	2.20

Table 3-1 adapted from Julian McCaull, "Building a Shorter Life," **Environment**, Vol. XIII (September, 1971), p. 9.

one team of researchers concluded: "Cadmium has probably more lethal potentialities than any of the other metals."[29]

There is evidence that cadmium is an important contributory cause of fatal hypertension (high blood pressure), a life-shortening disease estimated to afflict up to 23 million Americans. At some point in everyone's life, high blood pressure is normal. When we become scared, for example, our blood pressure goes up naturally, forcing freshly oxygenated blood into our brain at faster rates. However, at some point in many people's lives, high blood pressure ceases being a temporary reaction to stress and becomes a permanent condition. When this happens, high blood pressure must be considered a disease condition because it definitely shortens people's lives; those who have high blood pressure simply can't expect to live out their full measure of days.[30] Even if there are no other complications, high blood pressure by itself shortens life.

However, the attendant complications are frequently more serious than the high blood pressure itself. For example, under the increased blood pressure a blood vessel in the brain may break (a condition called cerebral hemorrhage or stroke or cardiovascular disaster of the central nervous system). On the other hand, high blood pressure will also accelerate the rate of progression in hardening of the arteries (a general condition called arteriosclerosis). High blood pressure also results in enlargement of the heart.

One way or another, high blood pressure is associated with the entire constellation of disease conditions that go under the general names "heart disease" or "cardiovascular disease." According to eminent physicians, heart disease is now considered an epidemic condition in the United States. Heart disease is now our biggest killer in the thirty-five- to fifty-five-year-old range; the heart disease rate among men and women in the prime of life is steadily rising. Heart disease not only is killing off our most experienced people, our elderly, but is now sapping the nation's managerial talent and its active work force at an accelerating rate.

As with so many important science discoveries, the evidence linking cadmium with heart disease came to light by accident.[31] A group of heart researchers in St. Louis, all one way or another connected with Dr. Henry Schroeder, made the initial discoveries. Early in the 1950s a new drug to lower people's blood pressure was coming into wide use. Called hydralazine, the new drug definitely worked for many people, yet no one knew *how* it worked. Then one day Dr. H. Mitchell Perry, Jr., noticed that hydralazine, stored on the shelf in bottles, was corroding the metal bottle caps. This corrosive action of the drug on metal suggested a series of experiments. Perry and other workers soon showed that hydralazine and numerous other drugs in common use (including penicillin and many antibiotics) share a common feature: They form a kind of molecule called a *chelate* (pronounced key-late). A chelate is a kind of molecule in which a metal lies near the molecular center; chlorophyll, with magnesium near the center, is a chelate. Chelating agents are chemicals that bind metals into chelates. Chelates ordinarily bind a metal so tightly that the metal becomes chemically and biologically inert thereafter. Chelating agents are now given to human beings as therapy for metal poisoning; the chelating agents hook onto the poisonous metals with very firm bonds and carry the toxic substances, now in inert form, out of the body in feces and urine.

The discovery that many drugs are chelating agents[32] was a very important one. For one thing, it led to a search for the metal (or metals) which might be causing high blood pressure in people. Soon other co-workers of Schroeder—among them Isabel Tipton and J. J. Balassa—had pinpointed cadmium as a metal frequently found in abnormally large amounts in the kidneys of people who die of high blood pressure.[33]

Other research during the 1960s revealed further evidence linking cadmium with high blood pressure. In rats (and to a lesser degree in people) therapeutic treatment with EDTA (a powerful chelating agent) reduces elevated blood pressure to normal. Furthermore, in people suffering from very severe hypertension, excessive levels of cadmium are excreted in the urine (50μg/liter as opposed to a "normal" value of 1μg/liter or less). In addition, hypertensive (high blood pressure) patients in the United States are thought to have increased levels of cadmium in their kidney tissues.

In rats, injecting cadmium ion (dissolved cadmium) into the bloodstream brings on acute hypertension. Feeding cadmium to mice at the level of 5 ppm in their drinking water brings on all the symptoms associated with human hypertension—large heart, changes in the blood vessels of the kidneys, high blood pressure, and an increase in atherosclerosis. When hypertension develops in rats, cadmium levels in the rats' kidneys are found to be very close to the levels now found in the kidneys of average Americans. When cadmium is removed from the rats' diet, their blood pressure eventually falls back to normal.[34]

In the mid-1960s Dr. Robert Carroll reported in the *Journal of the*

Table 3-2. Measurements of airborne cadmium in American cities, compared to death rates for diseases of the heart (excluding rheumatic diseases), 1959-1961.

City	Cadmium in air (micrograms of cadmium per cubic meter of air)	Heart disease death rate
Las Vegas, Nevada	0.000 micrograms (μg)	66.6
Eugene, Oregon	.000	83.5
Medford, Oregon	.001	96.7
Chattanooga, Tennessee	.001	87.6
Albuquerque, New Mexico	.001	71.7
Omaha, Nebraska	.002	101.0
Gary, Indiana	.003	103.0
Los Angeles, California	.004	95.4
Oklahoma City, Oklahoma	.006	75.8
Phoenix, Arizona	.006	82.5
Akron, Ohio	.006	95.2
Racine, Wisconsin	.006	109.0
Wilmington, Delaware	.006	115.4
Tucson, Arizona	.007	81.0
Youngstown, Ohio	.007	99.0
Cincinnati, Ohio	.007	111.2
Canton, Ohio	.009	103.9
Scranton, Pennsylvania	.011	135.2
New York, New York	.013	115.3
Columbus, Ohio	.014	103.2
Charleston, West Virginia	.018	115.8
Newark, New Jersey	.018	119.3
Indianapolis, Indiana	.019	102.5
Waterbury, Connecticut	.020	105.3
Bethlehem, Pennsylvania	.021	118.5
Philadelphia, Pennsylvania	.023	127.6
Allentown, Pennsylvania	.037	116.4
Chicago, Illinois	0.062	123.8

Table 3-2 adapted from Robert E. Carroll, "The Relationship of Cadmium in the Air to Cardiovascular Death Rates," Journal of the American Medical Association, Vol. CXCVIII (October 17, 1966), p. 178. Elsewhere Dr. Carroll notes that subsequent studies have shown a statistically significant correlation between airborne cadmium and cardiovascular death rates but the correlation has not proved as strong as the one shown in Table 3-2. See Robert E. Carroll, "Trace Element Pollution of the Air," in Delbert D. Hemphill, ed., Trace Substances in Environmental Health III (Columbia, Mo., University of

American Medical Association a strong correlation between levels of cadmium in urban air and human deaths from heart disease.[35] Table 3-2 shows the data which Dr. Carroll published.

There is now much evidence that cadmium can be measured as a contaminant in both the urban and rural air of the United States. There is also considerable published information showing cadmium in parts of the nation's water pipes, whether those pipes are galvanized iron, copper, or PVC plastic. Cadmium enters household water supplies in soft water areas rather than hard water areas, because soft water is frequently slightly acid compared to hard water. Generally speaking, the soft water areas of the country are the high rainfall coastal regions: the hard-water areas, generally speaking, are the drier inland regions. Ordinarily, surface waters fed by a lot of rain (which is soft) tend to be softer and more acidic than subsurface (so-called ground-) waters.[36]

Several experiments have shown that cadmium leaches out of household water pipes. Dr. Henry Schroeder placed short lengths of plastic pipe, copper pipe, and galvanized iron pipe into uncontaminated Vermont spring water and in twenty-four hours found 5 ppb cadmium in the water. Each type of pipe contributed cadmium. Five ppb is 50 percent of the amount set as "safe" for drinking water by the U.S. Public Health Service. PVC plastic pipe contains cadmium, which is added as a stabilizer to make the pipe stronger. Copper pipe contains cadmium because cadmium is added to copper to make it stronger for use in automobile radiators. Automobile radiators are recycled, and all the resulting recycled copper

---

Missouri, 1970), p. 228. Specific findings of subsequent studies remain unpublished so far as we know.

As we try to interpret Table 3-2, we should remember Dr. Carroll's words when he published the original information: "Any correlation coefficient must be interpreted with caution. The fact that two variables are strongly associated does not necessarily mean that one causes the other. A causal association of each with a third variable could produce the correlation. A possible explanation for the association of cadmium with heart disease might be that both tend to increase with city size and degree of industrialization. Heart disease does not correlate at all, however, with either total suspended particulates [in air] or benzene-soluble organics in air. Both of these measures of pollution are closely related to population size and industrialization and are accepted as indexes of general air pollution. In fact, except for zinc, cadmium is the only pollutant which shows a marked correlation with heart disease." (pp. 178-179) (some of which then gets made into water pipes) evidently contains some small amount of cadmium contamination.[37] Galvanized iron or galvanized steel water pipes contain cadmium as a contaminant in the zinc-galvanizing material. In 1911 the American Society for Testing Materials (ASTM) established standards for the level of cadmium (and also lead, iron, and aluminum) permissible in various grades of commercial zinc. Table 3–3 shows the latest revised ASTM standards (dated 1949) giving the maximum permissible contamination (in percent).

In several reports Dr. Schroeder has mentioned finding excessive levels of cadmium in tap water drawn from kitchen and bathroom faucets. Dr. Schroeder was sampling water at the consumer's direct source, and in 1971 he wrote, "We found more than one third of the tap water from private houses in the East unfit for human consumption according to government limits for cadmium."[38] The government limit for cadmium in drinking water has been set at 10 ppb (parts per billion) for drinking water safety. Schroeder has most recently reported finding 15 ppb—50 percent *above* the drinking water standard—in a "well advertised cola drink."[39]

Table 3-3. Allowable cadmium contamination in commercial grades of zinc.

| | | Contaminant | | Permissible total |
Grade of Zinc	Lead	Iron	Cadmium	not over (in percent)
Special High Grade *	0.006	0.005	0.004	0.010
High Grade *	0.07	0.02	0.07	0.10
Intermediate *	0.20	0.03	0.50	0.50
Brass Special	0.60	0.03	0.50	1.00
Selected *	0.80	0.04	0.75	1.25
Prime Western	1.60	0.08	No limit	No limit

* It shall be free of aluminum.

Source: Henry A. Schroeder, "Trace Metals and Chronic Diseases," **Advances in Internal Medicine**, Vol. VIII (1956), p. 277. Dr. Schroeder states that Prime Western grade zinc is used for galvanizing. He suggests that galvanized piping (in water supplies and in food-processing equipment) and brass (a combination of copper and zinc widely used in pipe fittings, especially for copper pipes) are "likely sources of the cadmium found to accumulate in human tissues." (pp. 277-78)

The Environmental Protection Agency recently tested 720 of the nation's surface waters, but unlike Dr. Schroeder, the EPA did not test water coming out of the tap; it tested raw water samples, water from the source of supply, before it had been put through the cadmium-containing distribution system of pipes. Still the EPA study revealed detectable levels of cadmium in 42 percent of all surface waters, and it found the raw water supplies of ten cities containing levels of cadmium higher than the "safety" limit of 10 ppb. The agency points out that it found officially unsafe levels of cadmium in only 1.4 percent of the nation's urban raw-water supplies.[40] This may sound acceptably low until we translate the percentage figure into numbers of people. Given the evidence linking cadmium with heart disease, and given the epidemic of heart disease that has been spreading throughout America in recent years, how can the 1,144,772 people who drink cadmium-polluted water feel safe? See Table 3–4 for a list of the cities with cadmium-polluted raw water supplies. And what of the statement that, "We found more than one third of the tap waters from private houses in the East unfit for human consumption according to government limits for cadmium"? There can be little doubt now that many of our domestic water supplies are not as safe as we once thought them to be. And from water, cadmium reaches not only direct consumers, but also indirect consumers, through the soil.

In some regions of the country, where irrigation farming is widely practiced, cadmium will reach food crops through soil which is contaminated by both air and water pollution. In other areas of the country, air pollution washed out by rain will make the most significant contribution to the soil. In almost all farming areas today—including many organic farms—cadmium reaches crops as a contaminant of phosphate fertilizers.

It has been estimated recently that dustfall in urban areas now contributes more than 2 grams of cadmium per year to each acre of the earth's surface. In rural areas, dustfall contributes an estimated 1 gram per acre per year. A normal application of phosphate fertilizer is estimated to contribute 1 gram of cadmium per acre.[41] Thus, both air pollution and direct soil pollution from fertilizers contribute cadmium to the American food supply. As a soil scientist with the federal Department of Agriculture recently wrote, "Cadmium is a metal of concern to agronomists because it is known to be absorbed readily by at least a number of plants, such as mint, vine, lettuce, parsnips, beets, peas and grain crops."[42]

72 NO WORLD WITHOUT END

Table 3-4. American cities with raw water supplies containing cadmium levels above the maximum "safe" limit * in 1970.

CITY	Amount of cadmium found in raw water supplies	Population in 1970 census
Birmingham, Alabama	12 ppb	301,000
Huntsville, Alabama	90 ppb	138,000
Hot Springs, Arkansas	20 ppb	35,631
Little Rock, Arkansas	18 ppb	132,486
East St. Louis, Illinois	16 ppb	69,804
Shreveport, Louisiana	16 ppb	182,000
Cape Girardeau, Missouri	20 ppb	30,914
St. Joseph, Missouri	20 ppb	72,932
Wilkes-Barre and Scranton, Pennsylvania	27 ppb	162,290
Pottsville, Pennsylvania	11 ppb	19,715
	Total:	1,144,772

* U.S. Public Health Service 1962 Drinking Water Standards set 10 ppb as the "safe" upper limit.

Source: U.S. Geological Survey, Reconnaissance of Selected Minor Elements in the Surface Waters of the United States — October, 1970 [Geological Survey Circular 643] (Washington, D.C.: U.S. Government Printing Office, 1970).

As Table 3–5 makes clear, U.S. phosphate fertilizer production in 1968 totaled nearly 11 billion pounds. What portion of this was cadmium contamination, it is difficult to say. Estimates of cadmium contamination in fertilizers range from 5 ppm to 170 ppm.[43] Using these outside limits, we can estimate that U.S. phosphate fertilizer production in 1968 contained somewhere between 55,000 pounds of cadmium and 1,870,000 pounds of cadmium. All this was intentionally put into the upper horizons of the earth's best soil (though not all of it inside U.S. borders since we exported a portion of our fertilizer production to foreign nations). And we now do this every year, year after year.

Commercial phosphate fertilizers are now standard soil additives for production of wheat, peanuts, tobacco, legumes, and grasses

Table 3-5.   U.S. phosphate fertilizer production, 1968.

Type of Fertilizer	U.S. production (in thousands of metric tons)
Direct phosphate rock application fertilizer	10.8
Fertilizers (from furnace phosphoric acid)	220.
Fertilizers (from wet-process phosphoric acid)	3,180.
Normal superphosphate	1,060.
Triple superphosphate	391.
Total	4,861.8

Table 3-5 adapted from Study of Critical Environmental Problems, **Man's Impact on the Global Environment** (Cambridge, Mass.: MIT Press, 1970), p. 271.   Figures adapted here are probably underestimates of actual total U.S. phosphate fertilizer production (see same source, p. 268).

___

throughout the United States. Many people do not realize that this heavy fertilization of the land by inorganic fertilizers, nitrates and phosphates is only a recent development. Fertilizer production has grown very rapidly and really only in the last thirty years. World fertilizer consumption is now increasing tremendously under pressures to feed more people; thus, the cadmium contribution to the earth's soil from this source seems certain to increase in the coming thirty years and beyond. As Figure 8–1 (in Chapter 8) shows, world fertilizer consumption is now doubling in annual volume every ten years.

Food crops in industrial countries now contain measurable quantities of cadmium, though the levels do not appear in most cases to be alarming by themselves. Table 3–1 shows the cadmium content of many food products. The two principal sources in the human diet are grains (excepting corn and rye) and shellfish (mollusks and crustacea: lobsters, shrimp, oysters, clams, and related creatures).[44] The Food and Drug Administration (FDA) in Washington hasn't yet set a "safe" level for cadmium in shellfish. At a Shellfish Sanitation Workshop in 1968, federal officials proposed a total allowable limit of 2 ppm on a combination of four elements in shellfish (the four elements being cadmium, chromium, lead, and mercury). The total concentration of these four together should not exceed 2 ppm, from a human health standpoint, officials said. Dr. Benjamin H. Pringle, of the federal Northeastern Water Quality Laboratory in Narragansett, Rhode Island, presented data to the workshop indicating that the average U.S. oyster is already contaminated above the suggested safety level. In fact, there is evidence that *from cadmium alone,* on the average U.S. oysters have already been polluted above the 2 ppm level.[45] As reported by the New York *Times,* Dr. Pringle "said there was 'considerable anxiety on the part of both the states and the industry' about the effect of such a guideline on the oyster industry since the average oyster already exceeded the limit."[46] There was so much anxiety, evidently, that the 1968 workshop threw up its hands: "Approval by the workshop would have been the first step in setting federal guidelines on what levels of contamination may be dangerous. But the workshop declined to take action on the findings and recommended only that further studies be made."

Depending on what foods one emphasizes, an American's daily diet of 3,100 Calories* can supply much more than 100 micrograms of cadmium each day or as little as 4 micrograms per day.[47] The cadmium-rich diet would include shellfish and grains high in cadmium. A high-cadmium diet might also include organ meats such as kidney and liver of cattle or hogs. A low-cadmium diet would consist of mostly vegetables and purified water.[48] If the drinking water isn't purified, it may contribute 5 to 10 percent as much cadmium as the food portion of the diet each day. Today the average American ingests a total ranging from 70 to 200 µg of cadmium daily from food and water.

The "safety" standards that the Food and Drug Administration established to protect the public from cadmium contamination has been seriously challenged by a report of the Swedish Natural Science Research Council. The "safety" standard is 5 ppm for food for humans; this means that in 1 million pounds of food if you find only 5 pounds of cadmium metal, you can consider the food "safe" to eat. The Swedish report, however, says this 5 ppm safety level "is

---

*"Calories" with a capital C refers to kilogram calories (kilocalories). One kilocalorie is enough heat energy to raise the temperature of one kilogram of water one degree Centigrade.

absolutely not founded on thorough toxicological investigations, but seems to be chosen completely at random. Considering available toxicological investigations in animals, and the Japanese analytical data available from the highly contaminated environment where the outbreak of [Itai-Itai] disease was observed, this recommended maximal level [5 ppm] appears to be set far too high."[49]

The outbreak of Itai-Itai disease occurred in Japan in the late 1960s; altogether a total of 223 Japanese died of this horrible malady, and 1,000 or more Japanese people suffered permanent disability. Itai-Itai disease developed in older people, mostly women, who had been exposed to a combination of cadmium and lead through rice grown with mine-waste-contaminated irrigation water. There is evidence that the effects of the poisonous metals were heightened by a vitamin D deficiency in the afflicted population. People who developed Itai-Itai disease first began to complain of aches in their bones; as the disease developed, it hurt too much to walk. The excruciating pain gave the sickness its name; itai-itai in Japanese means roughly "ouch-ouch." In advanced stages, Itai-Itai patients dare not even cough for fear of cracking their ribs, which, like all their bones, have become completely decalcified and as breakable as icicles. Twisted limbs, broken and split bones, and constant anguish are the symptoms of this unique industrial affliction.[50]

Recently an international group of ecologists, meeting at Williams College to assess global environmental problems, turned in this succinct summary of the cadmium situation:

A significant source of cadmium for environmental pollution is super-phosphates. Food grown on soil fertilized with super-phosphates—including coffee, tea, grapes, lemons, cereals and sugar—have been found to contain significant levels of cadmium. Eggs and seafoods may also contain cadmium. High levels of cadmium contamination have also been recorded for meat (beef kidney, 40 ppm) and game animals (grouse, 51 ppm). The processing of foodstuffs may occasionally increase cadmium concentrations. An additional source of cadmium contamination is from galvanized and black polyethylene pipes used in housing. Incineration of cadmium-coated hardware, cadmium-containing plastics and cadmium-painted materials release cadmium by volatilization directly to the atmo-

Table 3-6. Rate of growth of cadmium use in the United States, 1954 to 1968.

Year	U.S. cadmium consumption (in pounds)	Time it will take to double U.S. cadmium consumption (based on rate of increase, 1954–1968)
1954	7,499,000	
1955	10,684,000	
1956	12,711,000	
1957	10,966,000	
1958	8,242,000	
1959	11,589,000	
1960	10,126,000	
1961	10,184,000	
1962	12,146,000	
1963	11,482,000	
1964	9,365,000	
1965	10,431,000	
1966	14,780,000	
1967	11,561,000	
1968	13,328,000	
Total	165,094,000	15.6 years

Table 3-6 adapted from W. E. Davis and Associates, National Inventory of Sources and Emissions, Cadmium, Nickel, and Asbestos, Vol. I, Cadmium [Publication PB 192 250] (Springfield, Va.: U.S. Department of Commerce, National Technical Information Service, 1970), p. 40.

---

sphere and subsequently to water and land. Schroeder linked human deaths from hypertension with increased concentrations of cadmium in the kidneys. Similar results have been shown for other animals. Apparently, cadmium enters the food chain and moves rapidly to biological consumers, including man, where it accumulates because of a low rate of excretion.[51]

Having thus summarized the problem, the international group made the following recommendations:

1. A worldwide monitoring of cadmium pollution should be organized and funded.

2. "The present maximum permissible levels of cadmium in air, water, and food should be reduced."

Despite this straightforward suggestion from an eminent group of scientists, reducing cadmium contamination of the environment appears to be unlikely. If recent history can serve as guide, we have a rapidly growing cadmium pollution problem on our hands, one which will be difficult to control. Table 3-6 shows the rate of increase in use of cadmium in the United States, 1954 to 1968. The annual increase is about 4.58 percent per year, which means that our total annual use of this poisonous metal is doubling every fifteen years. By the end of this century our use of cadmium will have doubled *twice*. By the year 2000 Americans alone may be using (and dispersing into the environment) 50,000,000 pounds of cadmium each year, and that doesn't include cadmium released from fossil fuels or the use of cadmium-containing phosphate fertilizers.

> There is strong evidence that environmental contamination [by the poisonous metal, cadmium] has reached the point where the chronic effects already are occurring in people.[52]
>
> —JULIAN McCAULL (1971)

4

"The Past Is Prologue"

Lead

THE well-known American physician and research scientist Dr. Henry Schroeder recently said, "We cannot continue to dump lead into our air and our environment at the present rate without surely expecting widespread low-grade lead poisoning."[1] Another eminent scientist, Dr. Harriet Hardy, who recently retired from her position at the Massachusetts Institute of Technology, holds the opinion that many Americans are already experiencing the toxic effects of lead, that low-grade lead poisoning is now more common among us than most of our physicians suspect. Dr. Hardy reportedly thinks the symptoms of our chronic, low-level lead poisoning are nonspecific: a tired, run-down feeling, nervousness, frequent colds and other infections, apathy, lack of ambition, and other vague disturbances and disabilities.[2] Seven years ago Dr. Clair C. Patterson of the California Institute of Technology published his opinion that our air now contains so much lead that most urban and suburban Americans (130 million of them) are continually undergoing "severe chronic lead insult."[3]

Low-grade lead poisoning is extremely difficult to diagnose. Like cadmium, lead is a cumulative enzyme poison; at the microchemical level, lead interferes with several enzyme systems. Thus, as we might expect, lead poisoning symptoms can appear in several different parts of the body: the gastrointestinal tract, the central nervous system (spinal cord and brain), the heart, the kidneys, and the liver. As Anne and Paul Ehrlich said in 1970, "Chronic lead poisoning is unusually difficult to diagnose, and low-level effects could be quite common already."[4]

There is apparently no effective threshold level for lead poisoning, no "level below which no damage will occur." As has been shown to be true with radioactive materials and with methyl mercury, *any* amount of lead in the human body apparently does *some* damage. As one researcher said recently, "It is doubtful that there is a threshold for damage, and it is believed instead that damage to humans, especially to the central nervous system, is more or less proportional to the degree of exposure."[5] The average American today apparently ingests (through air, food, and water) 300 to 400 micrograms of lead daily.[6] Dr. Schroeder said in 1968, "We must look for some widespread and relatively new disorder of civilization as a consequence of sub-clinical lead toxicity, but what it might be, no one knows."[7]

All this, of course, raises difficult questions about the quality of life in the new industrial state. It would seem physically impossible for the human body undergoing continual assault from cumulative toxic materials, such as cadmium and lead, to perform up to its design potential; it would seem impossible for most people, living under modern industrial conditions, to feel a normal, full measure of health and vigor. Trying to assess the costs of air pollution, two analysts writing in *Science* in 1970 said, "We conjecture that the major benefit of pollution abatement will be found in a general increase in human happiness or improvement in the 'quality of life,' rather than in one of the specific, more easily measureable categories."[8] Perhaps one indicator of a general physical malaise among Americans is our remarkable consumption of aspirin (acetylsalicylic acid); Americans now consume 15 tons of aspirin each day.[9]

As far back as Egyptian civilization 7,000 years ago, human beings discovered that lead makes good glazes for pottery.[10] However, mining of lead didn't get underway on a major scale until about 800 B.C. We know these dates because scientists have discovered an excellent physical record of the world's historical mobilization of lead.[11] Year after year, atmospheric lead has been falling out of the sky onto the Arctic ice cap, providing modern science with a quantitative record of atmospheric fallout, going back thousands of years. As Figure 4–1 shows, lead fallout created a steady, low "background" level of lead in Arctic ice and snow until about

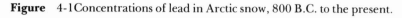
Figure 4-1Concentrations of lead in Arctic snow, 800 B.C. to the present.

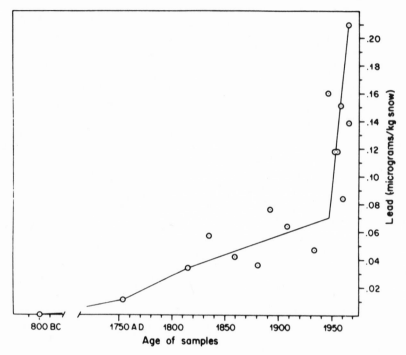

Figure 4-1 adapted from Robert A. Goyer and J. Julian Chisholm,
"Lead," in Douglas H.K. Lee, ed., *Metallic Contaminants and Human Health*
(New York and London: Academic Press, 1972), p. 61.

800 B.C., at which time human contributions to the earth's atmospheric lead became significant.[12] As is obvious from Figure 4–1,
modern industrial use of lead has been steeply rising since World
War II. Between 1904 and 1964 annual lead concentrations in
Greenland snow increased by a factor of 16.

Measured against natural background levels (the amounts of
lead which were falling out of the air onto Greenland snow before
people started mining lead commercially), our global lead pollution
is definitely significant. By 1750 lead in Greenland ice is elevated
above natural background levels by a factor of 25. By 1940 it is 175

times as high as the background, and by 1968 it is 500 times as high as the background.[13]

As far back as we can go in history, civilizations have been built on metals (and energy), and lead was one of the first metals abundantly available. At first, beginning about 2500 B.C., lead ore was mined on a small scale in the Middle East to extract valuable silver from it. With the invention of coinage about 650 B.C., the demand for silver increased rapidly, and so did the amount of lead ore which was being dug up and discarded. For every ton of silver mined, there were 300 to 400 tons of lead oxide left somewhere in tailings piles. Not surprisingly, people eventually found many uses for these tons of discarded, easily worked metal. Roofs, gutters, pipes for irrigation water and for plumbing, urns and cups for holding liquids and for cooking, glazes for pottery for the same purposes, a strong, durable base for paints and for cosmetics—all these economic uses were served well by lead. The history of lead's name shows how intimately lead and urban civilization have always been associated; the modern chemical symbol for lead is Pb, after the Roman (Latin) name for lead, which is *plumbum.* Modern medical doctors still refer to lead poisoning as plumbism. And today, when the essential pipes stop up and civilized life is threatened, we still call the plumber.

Understandably, one of the first by-products of early urban civilization was a great deal of human disease. People crowded into cities for the first time, and soon epidemics of bacterial and viral infections became evident. From the earliest days the connection between bad water and disease seems to have been recognized. So from the earliest civilized times people have been drinking water substitutes—chief among them, wines—which protect their drinkers against some of the worst waterborne diseases.

In ancient Greek civilization, winemaking had reached a high degree of art. In addition, Greeks used wine grapes as a popular sugar substitute because no sugarcane or sugar beets grew in their part of the world. One of the basic constituents of the ancient Greek diet was a thick grape syrup made by boiling down grape juice to about one-third its original volume. In the absence of refined sugar, this thick syrup came into widespread use for sweetening all sorts of dishes and drinks, as well as for preserving olives, plums, cherries, apples, pears, peaches, and quinces. In addition to preserving fruit with this grape syrup—called sapa by the ancient Greeks—sour wine was frequently sweetened with it. As we have noted, the ancient Greeks were good at making wine, and they made a lot of it.

Eventually, however, Roman civilization conquered the Greeks. Until this happened, the Romans hadn't used much wine because the necessary grapes hadn't been available to them. Now suddenly winemaking became possible and wine drinking became widespread among the Roman well-to-do. In addition to adopting Greek wine habits, the Romans adopted sapa, the grape syrup, though they changed its name to defrutum.

Now this defrutum is a rather acid substance, so it will tend to attack any metallic container into which it is placed. Although lower-class Romans might store their foods and drinks in plain pottery vessels, the well-to-do preferred the elegance of metallic containers. But bronze and copper would not do; the acidic defrutum (or the acidic wine) would leach copper from such vessels, producing the obvious effects of copper poisoning: a very bad taste followed by a great deal of vomiting. So Roman food vessels and drink containers came to be lined with lead or with a lead-tin combination called pewter.

The Roman author Pliny the Elder described how to make defrutum in his *Historia Naturalis*. "Leaden and not bronze pots should be used," Pliny advised.[14] Compared to copper (or bronze), lead is a much more toxic and much more subtle substance. Lead actually tastes good, and it doesn't bring on vomiting, or at least not until after a considerable delay. One lead compound (lead acetate) is called sugar of lead commercially because it is so sweet.

Defrutum was made in lead-lined pots by stirring grapes slowly over heat; the resulting syrup had to be high in lead content. Defrutum was also sweeter than the pure grade syrup and so, when wine soured, defrutum would be added to sweeten it again.

It has been calculated that the Roman aristocracy added lead to their diets in at least fourteen different ways. Defrutum was just one; the water used in winemaking might be another. The water would probably have been run off from a leaden roof and stored in lead-lined pots. The wine itself might be drunk from lead-lined goblets.

During Roman times, per capita use of lead reached 50 percent of what it is in America today. Romans used lead for pipes, cups, toys, coins, lids for containers and the containers themselves, sieves, solder, paint, cosmetics, and external medicines. Finally, lead was commonly alloyed with copper (lead bronze) and with tin (pewter) and used in many other ways.

In amounts exceeding 1,000 micrograms per day (1/28,000 of one avoirdupois ounce) lead can be extremely dangerous to people. Among its best-known effects are stillbirths, fetal deaths, miscarriages, and sterility, and these are the effects which evidently extinguished the Roman upper classes. During the first or second century A.D., the Roman aristocracy began failing to reproduce itself, and in three to six generations it was gone completely. Scholars have culled classical literature for references to the rate of reproduction among the ruling classes, and it seems to be a fact that very few aristocratic Roman couples successfully bore children. During each generation a large fraction of the nation's leadership apparently failed to leave any offspring, and in just a century or so the country's traditional leadership was gone, putting policy into the hands of a new group. No sooner had these new leaders emerged than the habits of upper-class society poisoned *them* with lead, and in time Rome fell. Of course, there were other factors involved in the fall of Rome, but lead is surely to be counted among the most important. It is worth pointing out that, among prevailing historical theories regarding the fall of Rome, only this one is based solidly on physical evidence. Poisonous levels of lead can be measured in the exhumed bones of Roman aristocrats. Your bones ordinarily contain 91 percent of your body's total burden of lead. This fact underscores the urgency of throwing off historic myths and starkly facing our situation: Except in the case of acute poisoning in the workplace (lead smelters), the Romans didn't recognize that lead was toxic and many of them poisoned themselves into oblivion. It *can* happen to whole civilizations. Those who do not learn from history are doomed to repeat it.

We Americans are now mobilizing, then dispersing into the environment, twice as much lead per person per year as the Romans were using at the height of their self-destruction.[15] The Romans were eating and drinking large quantities of lead; today we know better than to do that. But today we are *breathing* lead—something the Romans (outside the workplace, at least) didn't have to contend with. Our bodies absorb 5 to 10 percent of lead we eat, but we absorb 40 to 50 percent of the lead we breathe into our lungs.

As we shall see in Chapter 7, one particular portion of the American people is taking most of the health damage from lead—the children of the inner-city poor. Over the past five years it has become evident, however, that now the specter of lead poisoning—the kind that could lower an IQ 10 to 15 points or more and still not be suspected to exist—has invaded even the homes of the middle-class children of white America.[16]

There are repeated references in the literature to a pall of polluted air spreading sideways out of the inner cities. This cloud of lead and other poisons—most of it coming from automobile tailpipes—has evidently reached some sort of saturation point in the inner-city regions and is now increasing in area rather than density. Judging by soil pollution measurements and by measurements of lead in the blood of middle-class children, the suburbs can now no longer be considered safe, or healthy places, especially for kids.[17]

A handful of corporate officials made the decision to put lead into gasoline starting in 1923, to improve the burning qualities of the gas. The burning quality of gas is measured by an "octane rating" number. The higher the octane rating, the better the burning quality of the gasoline.

To obtain maximum octane at minimum price, the oil companies add tetraethyl lead to gasoline. Tetraethyl lead (TEL for short) is an organic compound of lead. Making the inorganic gray metal into an organic compound (hooking carbon and other atoms onto it) completely changes the material. Instead of being a soft gray metal, TEL is an oily yellowish liquid which evaporates. And instead of being very toxic like metallic lead, TEL is supertoxic—it has been measured and found to be 10 to 100 times more dangerous than the inorganic lead which decimated the Roman leadership.[18] TEL is so dangerous that it is difficult to buy any, even if you are a research scientist. And when you succeed in getting a half pint of TEL shipped to your laboratory, it comes encased in a steel container, which is itself crated inside a sturdy wooden box, which is sealed with a heavy lock. The whole crate is covered with warnings and precautions not to breathe any of the TEL fumes or get any on your skin because human skin absorbs TEL just the same as blotting paper does.[19]

Despite the extreme toxicity of this material, American automobile and petroleum corporations have been putting TEL into gasoline at steadily increasing rates since 1923. By conservative estimate, an average of 240,000 tons of lead (almost 500 million pounds) go into (and out of) American automobiles each year. Fortunately, when the TEL comes out the exhaust pipe a very large fraction of it has changed back into less-toxic inorganic lead. In the city of Los Angeles alone, 30,000 pounds of lead are dumped into the atmosphere each day.[20] As a direct result of these practices, lead can now be measured at unsafe levels in most of our urban air at all times. The established "safety" level in California and New York is 1.5 μg/m^3 of lead in air. This measurement means that if you filter 1 cubic meter (about 1 cubic yard) of air, you'll collect 1.5 micrograms of lead on the filter paper. An average person breathes 23 cubic meters of air each twenty-four hour day.

Most urban air today contains lead above the recognized safety level most of the time. The size of the city seems to make a difference. In New York City, the air averages 2.5 μg/m^3—67 percent higher than the safety standard.[21] In San Diego, California, the air sometimes reaches 8 μg/m^3 for as long as a week straight—433 percent higher than the standard.[22] At some locations near a freeway the air regularly contains 50 μg/m^3, and one measurement in Los Angeles has reached 71 μg/m^3. This last reading exceeded the safety limit by a factor of 46.[23] People who live at such locations will be physically assaulted by the air they breathe, a silent crime of violence.

During 1971 the federal Environmental Protection Agency (EPA) released a study showing that the lead in urban air has been rising in recent years. (Actually the EPA suppressed the report for more than a year until the editors of an environmental newspaper, *Clear Creek*, purloined a copy and leaked it to the national press.)[24] The suppressed EPA study showed significant lead increases ranging from 2 to 64 percent in the air of our major cities during the 1960s. For example, between the two periods of measurement (1961-1962 and 1968-1969), lead in Los Angeles air went from an average of 2.29 micrograms per cubic meter to 3.58 micrograms— an increase of 56 percent in seven years.[25] A different study showed the lead in San Diego air increasing at 5 percent per year— which means the lead would be double its present level in fourteen years if the current rate of increase continued.[26]

The first observable sign of lead poisoning is an inhibition of the body's ability to manufacture red blood cells. As we have seen, most of the body's metabolic activities depend on catalysts to initiate or speed chemical reactions. Many of the body's essential catalysts are large organic molecules, proteins. We've seen that these protein catalysts are called enzymes. When lead enters the body, it begins to circulate in the blood. Lead is soluble in human body fluid (called serum) whereas it is completely insoluble in water.[27] Because of its high solubility in body fluids, lead distributes itself through the entire human system, ultimately going into storage in the bones. By a process of competitive inhibition, lead knocks out essential metals, taking their places on enzymes but not doing the work that the essential metal (copper or iron or zinc or magnesium, for example) would have done. *Any* amount of lead in the body inhibits one particular enzyme called ALA dehydrase and prevents iron from reacting appropriately with heme in the formation of your blood's hemoglobin. At levels as low as 1 microgram of lead per 100 milliliters of blood, heme synthesis is interrupted measurably.[28] When lead interrupts heme synthesis, the raw materials that would have become hemoglobin are left floating in the bloodstream, and the kidney system begins cleaning them out and disposing of them with the urine. At this point these raw materials can be detected by sensitive chemical tests of the urine; this is one test doctors use to help find out how badly a person has been damaged by lead.[29]

It is known that ALA dehydrase performs other key functions in the body besides helping manufacture hemoglobin. It has functions in the brain, and particularly in the growing brain, the brains of humans in their first two years of life. Lead probably inhibits ALA dehydrase in the brain as it does in the blood. Since it is impossible to observe enzyme activity in living brain tissue in healthy individuals, this important aspect of lead toxicity remains only dimly understood.[30]

In addition to inhibiting ALA dehydrase activity and the activity of several other enzymes, lead has an attraction for the cell membrane that serves as skin for each of the body's 50 trillion cells. Ordinarily the cell membrane is a continual, orderly hubbub of chemicals being transported back and forth through it, oxygen and other nutrients being carried in, carbon dioxide and other waste materials being carried out. This process is called cell respiration, and its proper working depends on the ability of the cell membrane to allow only the right chemicals to pass through it. Lead changes the ability of the cell membrane to distinguish between various chemicals. For example, lead changes the amount of potassium that can enter a cell.[31] This can alter the electrolytic environment inside the cell. And so lead is called a systemic poison—its toxic activity is so generalized that it can attack and harm every bodily system.

The human body absorbs different amounts of lead, depending on how the lead enters the system. When we eat lead (which we do partly because some lead is natural in most all soils and partly because we're polluting our soils, especially those near cities),[32] our bodies absorb only about 5 to 10 percent of it. On the other hand, when we breathe airborne lead, a high percentage of it can remain in our lungs—40 to 50 percent or even more.[33]

Because of the different rates of absorption, airborne lead now contributes an estimated 50 percent to the body's daily lead intake. In other words, if we completely stop putting lead in gasoline, we will cut our exposure to lead by a factor of 2. (However, by reducing lead in gasoline, we will have to increase other gasoline additives which cause smog and which perhaps cause several lung diseases.)

As we write, the federal Environmental Protection Agency has announced it will try to reduce levels of lead in American gasoline by 65 percent by 1979, but this decision has been challenged in court by both environmentalists and industrialists.[34]

Airborne lead that doesn't enter our lungs may eventually enter our stomachs because lead falls out of the sky onto soil and onto the leaves of plants. One especially well-known plant—lettuce—concentrates lead into its leaves (providing perhaps 1 percent of the body's total lead intake over a year's time with an average American diet).[35] Many other plants are known to concentrate lead from the soil. As a rule, plants initially store lead in their roots; as contamination increases, the lead moves about inside the plant (translocates) into the stem and the leaves.[36] Investigators working for the Du-Pont Company (one of the nation's two large producers of lead for gasoline) point out that in many crops—such as corn—the lead stays in a part which humans don't consume. They fail to point out that commercial food processors frequently sell these portions of the crop to local farmers, who feed it to their cattle. Cows' milk and beef both add significant quantities of lead to the diet; in the United States meat alone is estimated to contribute 30 percent of our daily lead intake.[37]

The question of soil pollution from automobiles and other sources is not merely an academic matter. Between September 1968, and April 1970, forty horses and eighty sheep died of lead poisoning in Solano County, California. The horses were pastured 30 miles downwind from a lead smelter. Eventually the lead smelter was suspected to be the prime source of contamination. But

careful analysis of the isotopic distribution of the lead demonstrated that the smelter contributed only about 50 percent of the lethal doses and that fallout from auto exhausts contributed the other 50 percent.[38]

In December 1973, 2,700 people were reported to have "undue"
levels of lead in their blood because they lived within four miles of a
smelter in El Paso, Texas.[39] During the period 1969–1971 the
smelter, belonging to American Smelting and Refining (ASARCO),
reportedly emitted 1,116 tons of lead, 12 tons of cadmium and 1–2
tons of arsenic. Among the 2,700 people "at risk" from residing
four miles from the smelter, 131 children were reported to be suffering "nerve damage, and impairment of visual and perceptual
skills,"[40] according to newswriters quoting the National Center for
Disease Control in Atlanta.

Lead-containing pesticides have been in use for almost a century
now, though their use was cut back drastically after the discovery of
DDT's low-cost bug-killing power in about 1944. In some areas the
persistent lead has been retained in soils sprayed fifty years ago and
forage crops grown on old potato fields or in old fruit orchards can
still contain enough lead to kill cows.[41]

In many parts of the country lead is still used for water pipes. In
soft-water areas, such pipes may contribute a few micrograms each
day into a family's total lead budget. Lead was found in twenty-sev-
en samples of raw water at levels higher than the "safe" standards
during 1970, but this represented a small portion of the nation's
drinking water supplies, and we are assured by government authorities that we have no problem.[42] Still, lead piping remains in
use in "many older homes," and "it has long been felt that there is a
need to assay the health hazards to our population from this potential source of contamination."[43] Such an assay of the health hazards
has never been made.

The per capita dietary intake from soil (via food) and water
amounts to 300 to 400 micrograms of lead daily for the average
American.[44] Then, depending on where he or she lives, the exposure may be doubled by airborne lead breathed in and retained in
the lungs. Residents of the inner city breathe in the most, even
though they do the least driving.

' Lead from automobile exhausts—and 90 percent of all airborne
lead comes from this single source—hangs in the air for a very long
time because it is made up of very small particles, averaging less

than 1 micron in diameter.[45] It is a characteristic of such small particles (about one-tenth the diameter of an average single human cell) that they do not readily fall out of the sky in response to gravity, they do not collect on raindrops very well, and thus, they remain airborne for very long periods, months or even years. Another characteristic is their ability to penetrate into the delicate lowest reaches of human lungs and become trapped there. The lungs have protective mechanisms for removing most large airborne particles, but below 5 microns in diameter, and especially below 1 micron in diameter, the lungs retain a high percentage of particles which enter them.[46]

Probably the most important evidence about lead and public health was reported during 1970 but still hasn't gained wide circulation. Lead has now been shown to have a peculiarly destructive effect on the surface of human lungs. To understand this peculiar effect, we need to know a little about how the lungs work. The inner surface of the deep lung contains a kind of cell called a macrophage. The purpose of the macrophage is to remove foreign bodies. When a macrophage comes upon a foreign body, some piece of airborne trash you've breathed in, the macrophage throws itself upon the trash, envelopes it as a jellyfish envelopes its supper, and carries the foreign body upward out of the lungs, inching toward the exit. After the macrophage carries its burden so far, other mechanisms carry the foreign body upward with the sputum, and it eventually ends up in your stomach and you excrete it. Thus, you can think of macrophages as sweepers which keep the deepest, most delicate parts of your lungs clean.

During 1969 controlled experiments showed that lead reduces the number of macrophages in lungs. Both in rats and in human tissue, researchers found that increasing the airborne lead level always decreased the number of active macrophages in lungs.[47] Such an effect might cause all sorts of symptoms in people—increased pneumonia, for example, because healthy macrophages ordinarily carry away disease organisms along with the trash. Another possibility might be increased lung cancer because there is much evidence that some foreign bodies, if not removed, cause lungs to react by building a tumor around the foreign object, and tumors can turn malignant and cancerous by unknown mechanisms. Thus, your macrophages protect your lungs against diverse diseases, and atmospheric lead increases the rates of other diseases besides direct lead poisoning. Probably the widespread ailments bronchitis and asthma, which are now understood to be the forerunners of terminal emphysema, can be traced in part to the lead (and cadmium) in our urban air.[48] The researchers who studied this macrophage phenomenon noticed that after prolonged exposure (longer than one year), taking the lead away didn't cause all the normal macrophages to reappear. Some *permanent* damage to the lung's ability to manufacture, or to support, macrophages seems to be involved. Permanent lung damage from only a year's exposure must be sobering news to many people. These findings occur at levels of 150 $\mu g/m^3$, which is below the allowable standard for lead in the American workplace $(200 \mu g/m^3)$. Further research must be undertaken to assess the seriousness of this aspect of the lead public health hazard. In the meantime, the industrial standard should definitely be tightened to protect workers' health.

Lead accumulates in the body over time. The body does have mechanisms for excreting lead (macrophages are part of one such mechanism), but these mechanisms evolved during a time when lead wasn't found at high levels in the environment (least of all in the air). According to measurements and calculations by Tsaihwa Chow and his colleagues, average modern urban air contains more lead than primitive, pure air by a factor of 113,000 to 333,000.[49] For this reason, as in the case of cadmium, lead builds up in people because human bodies haven't had time to evolve means for handling large amounts of this poison. The half-life of lead in humans is on the order of two years. We simply can't get rid of it fast enough to maintain full health at the cellular level under modern urban conditions.

When researchers artificially contaminate food for mice and rats to raise their body lead levels up into the range in which most Americans are now contaminated, the mice and rats show numerous signs of stress. For one thing, they die younger than lead-free mice and rats; for another, they die more frequently from infectious diseases such as pneumonia—adding strength to the evidence that macrophage defense systems and other defenses may be harmed by lead, allowing disease agents to persist in the lungs and establish themselves. Mice and rats with human levels of lead contamination age more visibly as they get older and they continually lose weight. (Among the Romans there was a persistent belief that fat women bore more children than thin ones; this may actually have been true in Rome because lead causes people to lose weight and it sterilizes them at the same time.) Lastly, lead at the same levels now found in people causes rats and mice to become sterile. Laboratory experiments have shown that in two to three generations an entire litter will die out.[50]

Do American women have more miscarriages, defective births, and spontaneous abortions than they used to? The question cannot be answered. We have sought the answer from numerous sources (the federal government, university medical schools and their libraries, professional associations, all standard medical and biochemical abstracting services, and insurance statistics), and we have been told repeatedly that no adequate records are kept on such matters. The whole field is virtually a blank. At least three data collection projects are now under way, but they have almost no publishable results yet, and according to some professionals, none of the studies currently under way can have much significance at the national level because doctors in most states still aren't required to report birth information in any detail. For example, in our state, New Mexico, doctors have only in the past four years been asked to report birth defects on the birth-register form, and the space for reporting birth defects consists of a small printed box suitable only for an X or a check mark if there is any sort of birth defect. No space for any description has been provided, and no law requires a physician to report anything at all if he or she doesn't want to or doesn't have time to. We understand from teratologists (specialists in birth defects) that the picture in other states is as bleak and unpromising.[51] As a nation we are still some years away from knowing accurately how many birth defects are occurring and how many are environmentally induced. Such an information gap can have very serious long-range consequences.

There are other aspects of lead contamination, besides birth defects and miscarriages and sterility, which should give us serious pause. Lead (like cadmium, as we have seen) has been implicated in human and animal heart disease. In particular, the thickening of blood vessel walls—atherosclerosis—has been pinpointed as resulting from lead pollution. Atherosclerosis frequently leads to heart attacks. In non-human species—particularly water birds—there is very strong, direct evidence linking lead pollution to heart attacks (called, in this case, myocardial infarction).[52] Water birds, such as ducks, wild geese, and wild swans, frequently eat lead shot that hunters have blasted into the sky from shotguns. If the shot doesn't lodge in the body of a flying bird, it lands nearby, often in the shallow water where waterfowl feed. Birds don't have stomachs; they have gizzards. In a bird's gizzard, food is ground up by the action of pebbles which the bird must eat along with its food. In eating pebbles from the bottom of shallow areas, ducks and other water birds frequently eat lead shot. The lead shot gets ground up by pebbles and dissolved by digestive juices in the gizzard, and the lead is distributed through the bird's bloodstream. Hunters blast 6,000 tons of lead into the U.S. sky annually. An estimated 2 million to 5 million water birds die in the United States each year from lead poisoning.

The point here is that numerous studies have shown, through autopsies performed on dead birds, that they died of lead poisoning and that the immediate causes of death were kidney damage and myocardial infarction—literally a loss of blood to the main heart muscle, which is the same thing as coronary thrombosis, or simply heart attack. If lead causes atherosclerosis and kidney disease and heart attacks in ducks and swans and geese, and if it is known to cause atherosclerosis in workmen at lead-battery factories,[53] then we can certainly be justifiably suspicious of a connection between the rising levels of lead in our air and the rising frequency of heart disease, which is killing a larger and younger population of Americans each passing year. And we can justifiably establish strict controls.

Lead has been implicated for a long time with kidney ailments and kidney disease.[54] The kidneys are filters that keep the bloodstream clean. They remove contaminants and pass them from the body with the urine. Kidney damage has been known for a long time to contribute to hypertension—high blood pressure. Since hypertension can contribute to many forms of heart disease, any damage that lead might do to the kidney is doubly worrisome because it will probably complicate any existing picture of heart trouble.[55] There is evidence that lead workers in battery factories suffer frequently from stroke—the breaking of a blood vessel in the brain.[56] Stroke is linked to high blood pressure.

And so we come full circle and see that lead (and also cadmium) is a significant factor that complicates a whole constellation of diseases which as a group are shortening the lives of the majority of us. For the majority, there is at least a theoretical solution to the lead part of the problem: stop putting lead in gasoline. This would immediately cut in half the normal exposure to lead, the airborne half that we breathe in directly.[57] It would also, in the long run, prevent further soil and food pollution from airborne lead fallout.

However, this solution will not touch the most acute portion of the nation's lead poisoning problem: the inner-city children, nearly half a million of whom are now suspected of having mild to moderate cases of lead poisoning. These nearly half million children are to be sacrificed, it seems. As we shall see in a later chapter, we know they are sick and getting worse, but as a nation we can apparently conceive of no public health measures adequate to save them.

> The general opinion today is that with our present rate of lead pollution, we will incur diseases related to lead toxicity.[58]
> —THE INSTITUTE OF ECOLOGY (1972)

5

The Critical Importance of Delay

Mercury

THE mercury crisis of 1970 made it perfectly clear that the nation's elaborate environmental protection apparatus didn't work.[1] In an editorial in September, 1971, the prestigious *American Journal of Public Health* asked the hard question, "In short, why did our environmental protection system fail?" The *Journal* went on to suggest some key reasons. "Inadequate resources, incompetence, misjudged priorities, a general apathy toward the environment can all be cited," said the *Journal*.[2] We find all these reasons persuasive. But another important reason hasn't been mentioned. The nation's environmental protection apparatus didn't work because we have continually failed to appreciate the importance of delays in environmental systems. To make our meaning clear on this, we must go back into the mercury story to see how delays played a critical part in creating the presently intensifying mercury problem.

Americans (and other industrialized peoples) started contaminating the environment with poisonous mercury around 1820, when coal burning first gained popularity in this country. (Mercury is a normal constituent of coal, and it is released into the biosphere when the coal is burned.) Following a period of delay lasting more than 130 years, environmental mercury poisoning (from polluted fish) finally caused recorded human deaths in 1953. That year Japanese medical authorities drew attention to an outbreak of a bizarre and deadly illness among families of fishermen living along Minamata Bay.[3] People reported being dizzy at first; then they no-

94

ticed they were having difficulty speaking. As the illness progressed, the people said their vision became constricted, as if they were looking through long tunnels, seeing only pinpoints of light in the center of everything. The disease progressed to permanent blindness, then drunken staggering, then to total loss of control over arms and legs (it is called *ataxia*), then to raving madness, followed finally by merciful coma and death. Cats came down with the disease and ran screeching through the village, and some of them ran into the sea, drowning in anguished madness. Even crows caught the disease; some of them flew crazily and fell out of the sky repeatedly until they died. Because of the cats and crows, the disease was soon traced to poisoned fish, but beyond that health authorities couldn't explain a thing. For seven years after the first deaths no one knew it was mercury poisoning.[4] This was the second important delay.

The third important delay stretches from 1953 through 1970—seventeen years during which U.S. and worldwide health authorities failed to respond to mounting evidence that mercury poisoning was beginning to turn up regularly among human populations. In 1956, peasants in Iraq were given mercury-treated seed for planting.[5] The peasants were warned that the seed was coated with poisonous mercury. The seed was coated with pink dye to indicate danger, but the dye was washable, whereas the poison itself was not. The mercury had been put on the seeds to protect against microorganisms during the first few days in the ground. No one is ever supposed to eat mercury-treated seed. But the Iraqi peasants fed the poisonous seed to their chickens and watched and waited. When the chickens remained healthy for more than a couple of weeks, the peasants began baking the seeds into their own bread and eating it themselves. Unfortunately, the symptoms of mercury poisoning don't show up in just a couple of weeks; there is a delay of one to three months between eating a poisonous quantity of mercury and the onset of symptoms. And this, of course, was the fourth important example of delay. By the time Iraqi chickens began to show signs of mercury poisoning the Iraqi people were already poisoned. Their own debilitating illnesses showed up a couple of months later. The corridors of the hospitals in Baghdad swelled with hundreds, then thousands of blind, mad victims of mercury poisoning, many of them irreversibly brain-damaged, restrained in straitjackets or tied to their beds, moaning and sobbing and laughing hysterically, waiting for coma or death to come on. Because of the critical delay involved in the onset of mercury poisoning, by the time symptoms appear it is too late; irreversible damage has already been done.

In 1960 a second, larger outbreak in Iraq followed the pattern of the first. Then in 1965 a second outbreak in Japan, this time on the island of Niigata, came to light. Here again, people were eating polluted fish. In 1966 a group of peasants in Guatemala—hundreds of people—were reportedly poisoned by mercury-treated seed, just like the people in Iraq.[6] Between 1960 and 1970 more than 450 people in several countries were reported poisoned by mercury.[7] Yet throughout this decade complacent American health authorities failed to see the picture developing.

More subtle—and ultimately more ominous—evidence besides poisoned people kept turning up throughout the decade of the 1960s. In the late 1950s Swedish bird-watchers had noticed fewer and fewer birds throughout the Swedish countryside. By the early 1960s careful scientific detective work had pinpointed mercury-treated seed as the agent killing the wild birds. During planting time, thousands of birds were eating poisoned seed; in those years Swedish farmers treated 80 percent of their seed with mercury. This agricultural practice had begun on a massive scale in Sweden (and the United States) during World War II.[8]

Interest in the mercury problem intensified considerably in 1966, when a leading Swedish researcher presented evidence that Swedish fish were dangerously contaminated with mercury. This was a new problem entirely, evidently not related to the birds. Could the tragedy of Minamata Bay be repeated in Sweden? Now events moved swiftly as Swedish and Japanese researchers probed more deeply into the complicated chemistry of mercury pollution. During 1966 and 1967 it was established that most of the mercury found in fish took the chemical form of methyl mercury—the most deadly form mercury can take.[9] Compared to plain inorganic metallic mercury, methyl mercury (which is an organic molecule) is 100 times more toxic.[10] It was methyl mercury that killed and maimed the Iraqi people and the Guatemalans; to everyone's surprise and then horror it was also 50 to 95 percent methyl mercury that was being discovered in the fish of Minamata and Niigata, and now Sweden.

In the period 1967 through 1970, teams of Japanese, Swedish,

British, and American researchers began the long task of examining the behavior of mercury in natural ecosystems, trying to find out where the particularly poisonous organic molecule (methyl mercury) was coming from. In the case of agricultural seed the methyl mercury had been added by agrichemists on purpose, but very few industrial operations dump methyl mercury into rivers or lakes. They more often dump plain inorganic metallic mercury or an inorganic mercury salt (such as mercuric chloride). How was all the superpoisonous methyl mercury getting into the fish?

Following evidence provided by Japanese and Swedish researchers, the American investigator Dr. J. M. Wood of the University of Illinois provided the first conclusive answer to this question. Dr. Wood revealed that microorganisms in soil and in underwater sediment continually convert ordinary mercury to its most toxic form and then excrete it from their own cells.[11] At first scientists thought that creatures could perform this methyl conversion only in the absence of oxygen, but more recent evidence indicates that methylation of mercury can take place even more rapidly when oxygen is present.[12] In addition, a recent survey shows that every surface water system in the U.S. contains organisms that can convert inorganic mercury to methyl mercury.[13] More recently, organisms that can demethylate mercury have been identified in sediments.[14] Obviously the aquatic biochemistry of mercury is complex.

Throughout 1969 and 1970 the evidence continued piling up. Late in 1969 the Canadian province of Alberta discovered that wild pheasant contained excessive mercury, and Alberta immediately cut short its hunting season that year. California health authorities reacted differently. First they pointedly ignored the Canadian warning until after the California hunting season was over. Then they quietly announced that they had found levels of mercury in California game birds four to five times higher than those in Alberta. It has since been reported that California authorities had known about the mercury problem in their game birds for five years but had suppressed the information.[15] It later turned out that California had set the tone for the entire American response to the mercury problem.

A freak accident in late 1969 brought mercury pollution dramatically to public attention in early 1970. Three children in a black family in Alamogordo, New Mexico, began showing typical mercury symptoms: blindness, staggering and falling down, incoherent speech, finally raving madness and coma. The children had eaten pork from a hog which had been fed mercury-poisoned grain by mistake. The pork contained 28 ppm methyl mercury. Early in January 1970, three months after the children had begun eating the pork, mercury symptoms began appearing and within a few weeks the childrens' minds were gone.[16] Still American health authorities delayed any national action.

Then in March 1970, a Norwegian graduate student at a Canadian University, Norvald Fimreite, announced that he had measured mercury up to 7 ppm in fish taken from Lake St. Clair, an important commercial fishing lake which feeds into the Great Lakes. Almost immediately Canadian authorities closed the lake, and now U.S. authorities had to act.[17] The seventeen-year bureaucratic delay period was over, and the American public health protection agencies began ponderously gearing up to measure the magnitude of the problem. Now another delay occurred while authorities gathered essential data and rationalized their public health positions. We are in this particular delay period as we write; it promises to extend into the 1980s.

Throughout the summer of 1970 U.S. government officials kept trying to minimize the problem. Officials as high as the President's Cabinet made dramatic headline-grabbing announcements that the situation was under control. Secretary of the Interior Walter Hickel at one point claimed that 86 percent of all mercury emissions to the environment had been stopped.[18] This was nowhere near true.

While federal officials did their best to keep the truth from getting out, other people were saying the situation was much worse than first suspected. These other people were not politicians but scientists, and their words began to take on an urgency rarely sensed in scientific literature. An editorial in *Science* magazine, official publication of the American Association for the Advancement of Science, said:

> Recently it has become clear that compounds of mercury present a substantial hazard. Of particular significance is methyl mercury, a highly toxic substance that causes neurological damage, produces chromosomal aberrations, and has teratogenic effects. . . . Further investigations have confirmed the existence of a major environmental

problem traceable to the dumping of large amounts of mercury-containing liquid wastes. This discovery comes as a surprise to most scientists and apparently to federal authorities. However, there was ample reason for looking for such a phenomenon. Episodes in Sweden and Japan had pointed to dangers. . . .[19]

Federal authorities tried to minimize not only the severity of the mercury problem but the extent of the problem as well. The government announced in July 1970, that mercury pollution affected only fourteen states.[20] However, Gladwin Hill of the New York *Times* published his own survey of the states sixty days later, revealing serious mercury pollution in thirty-three states. Hill reported that the remaining seventeen states didn't know whether they had a mercury problem because they hadn't looked adequately for signs.[21]

Then, in December 1970, mercury was discovered across the nation in swordfish and then in tuna fish and then not in tuna fish. Swordfish was banned from interstate commerce by the federal Food and Drug Administration (FDA), but 96 percent of the tuna fish tested were finally found "safe" and tuna remains for sale.[22] All of this poisoned-fish publicity, however, had raised a nagging question in the public mind: How much mercury is "safe"?

In an attempt to answer that key question, the President's newly created Environmental Protection Agency (EPA) commissioned an in-house study of mercury pollution, including an exhaustive review of safety standards. But when that study came in, EPA Administrator William Ruckelshaus immediately suppressed the report. For six months the suppressed report lay under wraps at the EPA; then columnist Jack Anderson broke the story this way:

The Environmental Protection Agency (EPA) which is supposed to alert the public to environmental dangers, has suppressed an alarming, 123-page report on the mercury menace. . . .

Conservationists have begged in vain to obtain release of the devastating report, which is entitled "Position Document Mercury." Every copy is marked "Not for distribution or quotation."

The "Emergency" study . . . was prepared by an EPA

> task force under the chairmanship of Victor Lambou,
> considered one of the nation's most eminent experts on
> mercury poisoning.
>   . . . The report has been suppressed, say insiders, be-
> cause the recommendations are too tough. The task force
> calls for an immediate end to almost all mercury release
> into the air, land or water on the theory that it is better to
> inconvenience industry than to poison coming genera-
> tions.[23]

After Anderson broke the story, we obtained a bootleg copy of
the EPA report.[24] The suppressed report tells a story that the
American people should know. According to the document, the
national mercury pollution problem is more serious than anyone
had thought. Anyone eating substantial quantities of fresh fish
from American waters today risks an appreciable hazard of mercury poisoning. The quantities of mercury in the fish in many surface
waters of the United States approach the quantities found in the
fish which poisoned the people in Japan. The main difference
seems to be that most Americans eat much less fish than the people
at Minamata and Niigata ate. As Dr. Neville Grant recently wrote,
"In conclusion, it can be said that no widespread epidemic of mercury poisoning is underway in the U.S., perhaps in large part because U.S. eating habits do not include large amounts of fish."[25]
The small percentage of Americans who *do* eat a lot of fish—probably mostly poor people—are definitely risking brain damage today.
The risk of brain damage in their unborn children is even greater.
    The U.S. "safety" standard for mercury was established in July
1969, by the Food and Drug Administration. However the standard
was set in an unusual manner. An official of the FDA told a symposium of scientists in 1971 that the safety standard had originally been
set at 0.5 ppm in fish because at that time the FDA's testing equipment couldn't detect lower concentrations than that. The present
U.S. safety standard was thus established not on the basis of safety,
but on the basis of available technology.[26] (The same official told the
symposium that the FDA has since acquired new expertise and
equipment and can now measure considerably smaller amounts.
Thus, it has now become technically, if not politically, possible for
the FDA to tighten the standard. Mercury can now be measured
routinely in the parts per trillion range.)

The EPA's own report—which is still being suppressed inside the EPA so far as we can determine[27]—clearly shows that the FDA's "safety guideline" should be considered loose and permissive. The report says:

> . . . It can be shown that the intake of mercury from all sources at levels considered acceptable can add up to a total intake which is unacceptably high. Therefore we must reassess the need for more stringent standards for mercury in air and drinking water, as well as from other sources, including fish.[28]

The report calculates, from actual measurements of brain-damaged human beings, that a person's daily intake of mercury cannot safely exceed 0.03 milligrams.[29] A milligram is a thousandth of a gram, and there are 454 grams in 1 pound. The measure 0.03 milligrams is equal to 30 micrograms. Eating more than 30 micrograms of methyl mercury per day is dangerous.

Based on recent measurements of airborne mercury vapor in urban environments, the EPA team concludes that normal breathing in urban air may be an important source of mercury in some individuals.[30] In addition, our drinking water and other beverages, including beer and soft drinks, may contribute significant amounts to our total mercury intake.[31] Other foods besides fish certainly contribute additional mercury to our diets. How much of this is methyl mercury is not known, but it is probably a small fraction.

The report points out that the maximum permissible daily intake (0.03 mg per day) would be "reached *without any fish consumption* if presently considered standards for mercury in drinking water and air were reached but not exceeded. If *any* fish were consumed, the daily acceptable intake would be exceeded."[32]

Approaching the human diet from another perspective, the report notes that a person eating 1 pound per week of fish contaminated at 0.5 ppm (the top limit of the "safe" scale), *eating or drinking nothing else*, would ingest, as a daily average, 30 micrograms of mercury, the maximum thought to be "safe." On the basis of these figures, the report asks whether our limit of 0.5 ppm in fish shouldn't be made stricter.[33]

After these elaborate arguments, the suppressed report then goes on to note that even its strictest safety calculations may not actually protect the public adequately because "brain lesions may occur at lower levels" than the "safe" 30 microgram daily intake figure.[34]

Indeed, up to now all FDA and EPA thinking on safety standards has failed to protect the public adequately because officials have always assumed that the person being protected was a fifty-year-old 154-pound white male in good health. None of our official thinking on safety has yet taken into consideration the likelihood of genetic effects from mercury, the likelihood of birth defects from mercury, the possibility of cancer, the special sensitivity of old people and children and especially babies and fetuses to mercury, and the additive (or, actually, multiplier) effects of mercury plus cadmium plus lead plus a hundred or a thousand other chemicals in small quantities, plus malnutrition, plus illness, plus radiation exposure.

Canadian health authorities (who have set the same "interim guideline" as the United States—0.5 ppm) have recently questioned the actual safety offered by such a standard;[35] and in the United States Dr. Barry Commoner has publicly stated that the FDA safety standard may not be safe at all.[36]

On the other hand, some scientists have been arguing for a more permissive mercury standard, on the basis that it's methyl mercury we're really worried about and we've set the standard on the basis of total mercury.[37] As we've seen, the mercury in fish is almost always more than 50 percent methyl mercury and it can run to almost 100 percent methyl mercury. Moreover, the most complete review of the mercury literature concluded recently said, ". . . extreme caution should be exercised in considering revision of the FDA's 0.5 ppm guideline."[38] Extreme caution.

There is likely to be a very long delay before we really know whether our current mercury standard is adequate for protecting all or even most of the people of the nation. Since mercury demolishes nerve cells in the brain and since mercury actually concentrates in the blood of children while they are growing inside their mothers prior to birth, mental damage in newborn children would be one of the results to expect when a group of people eat a lot of fish.[39] Investigators from the federal Center for Disease Control in Atlanta, Georgia, flew to the Pribilof Islands in Alaska three years ago to investigate the health of a group of natives there. The investigators found that almost all of the 500 families on the island ate fish and seal meat frequently.[40] They also found that both the fish and the seal contained an average level of mercury exceeding the FDA "safety" standard. Furthermore, they found that the average level of mercury in the people's hair averaged 4.45 ppm (parts per million)—an average three times as high as the typical U.S. citizen.

Significantly, the investigators also reported finding mental retardation in at least one individual in each of 30 percent of the families they investigated. This is a strikingly high incidence of mental retardation. The investigators said they could not implicate mercury as *the* cause of the mental retardation because they didn't have sufficient evidence. They were right in this. Establishing the cause of this particular epidemic of mental retardation would require complete biochemical and physical examination (and probably postmortem autopsy), as well as a detailed knowledge of dietary and medical histories. Mercury does cause a particular pattern of brain lesions—among them, complete destruction of granular cells in the cerebellum region of the brain—but several other illnesses can cause an identical pattern.[41] Despite these intellectual cautions, the brain-damaged Pribilof Islanders stand like a grim question: What are we doing to ourselves *really?* The answer is: We do not know. We have embarked on a massive uncontrolled biological experiment upon ourselves; we must expect the results of this experiment to be *delayed* in coming in. Until then we can't know what we've done or what we're doing.

Dr. Thomas Clarkson reports, after studying human populations in Peru and Samoa, that reliance on oceanic fish as a mainstay of the diet may produce mercury damage to the brain.[42] A recently-reported study has indicated that violent crimes among Ojibway Indians in Canada seem to be correlated to elevated tissue levels of mercury.[43] In brain damage, delay takes on critical significance. Brain damage from mercury poisoning may not show up in a poisoned individual for years and years. If the rate of poisoning is relatively steady but slow, as in the case of someone eating a pound or two of mildly polluted fish per week, mercury damage might not show up until the person was well into middle age. Dr. Neville Grant, a professor of medicine at Washington University in St. Louis, has described the long-term delayed effects of eating poisoned fish:

> From the nature of the injury in the nerve tissue [in the brain] caused by mercury poisoning, it is clear that the

absence of symptoms does not mean the absence of damage. The damage may go on out of sight for some years and may be of great significance as the poisoned person ages. The malfunction of the nerve cells damaged by mercury may at first be compensated for by viable neighbors, but as these normal neighbor cells are removed by the aging process, neurological abnormalities could become markedly enhanced.[44]

Dr. Grant was describing a condition known as presenile dementia, the early onset of senility, a general waning of the intellectual powers at an earlier-than-normal age. This kind of nonspecific debility could be widespread throughout large populations and still might never be suspected to exist. It could sap a nation's vigor and resolve.

Another form of delayed damage from mercury may show up 100 years after a person has been poisoned, in the fourth or fifth generation down the line. Mercury has definitely been established as a cause of chromosome changes in people. Our genes and chromosomes determine how our children will be formed physically and mentally; genes and chromosomes are the specific, chemical mechanisms of inheritance. In experiments with plants, insects, and people, mercury has been shown to disrupt genetic mechanisms.[45] In plants and insects, mercury has been shown to cause production of an extra chromosome. In human beings the production of one extra chromosome leads to serious birth defects: so-called mongoloid idiocy (Down's syndrome) and Klinefelter's syndrome (which is characterized by atrophied testes, the absence of living spermatozoa in the semen and sometimes excessive development of male mammary glands).

Mercury has not been definitely shown to produce an abnormal number of chromosomes in people. (The necessary experiments are out of the question.) It *has* been demonstrated, however, that elevated levels of mercury in human blood are associated with an abnormal number of broken chromosomes. Researchers took blood samples from a group of Scandinavian people who had been eating fish polluted at the same levels we are finding in some fish in most of the states today. These researchers found "statistically significant" numbers of broken chromosomes in the people, indicating that mercury may cause serious genetic damage in humans.[46] It now remains for medical researchers to determine what specific kinds of genetic damage to look for. This will be a big job and it will take years (even assuming the funds are made available). Most recently, we have a newspaper report that Dr. Lawrence Hecker of the University of Michigan School of Public Health has found chromosome damage associated with high levels of mercury in a population of Yanomama Indians in northern Brazil.[47]

Reporting findings of subtle nerve damage in mice from low levels of mercury contamination, Minnesota Medical School researchers summed up, saying, "It seems likely that many people with minor symptoms or subclinical damage have gone undetected."[48]

In the meantime, people want to know how much fish is safe to eat. Probably the best answer to this question comes from Dr. Norvald Fimreite, the researcher who first alerted the Canadian government to the mercury in Lake St. Clair. Dr. Fimreite says, "Asking how much mercury is safe to eat is like asking what speed is it safe to drive in a car."[49] If you want complete safety in your car, don't start the engine; if you want to be completely safe from mercury pollution, don't eat any more mercury than you absolutely have to.

However, you do have to eat, and fish is an inexpensive source of animal protein which is exceptionally nutritious. You should therefore eat fish, but in moderation. Tuna fish a couple of times a week is probably safe, but more than that would certainly be questionable.

In the case of swordfish, avoid it—on the average, it contains four to six times as much mercury as tuna fish. Dr. Bruce McDuffie has announced finding mercury five times higher than normal in the blood of a group of Americans who were eating a special diet high in swordfish.[50] No physical damage is evident in these people today, but we have seen the importance of delays in mercury (and other pollutants), so we shouldn't conclude too early that no damage has been done.

At the University of Wisconsin Medical School, researchers fed sixteen cats a strict tuna fish diet for seven months straight. The tuna was contaminated at levels ranging from 0.3 ppm to 0.5 ppm. In other words, it was tuna fish right off the shelf, contaminated (as almost all tuna is) at levels termed "safe" by the FDA.

At the end of seven months, three of the cats showed nervous disorders characteristic of gross mercury poisoning. Upon autopsy, all sixteen cats exhibited physiological changes, including damage to brain and liver typical of mercury poisoning. Reporting the findings, Dr. Louis W. Chang stressed that an adult male human would have to eat 400 pounds of tuna fish to reach the same levels the cats reached; that's a little over 1 pound a day for a year. However, he stressed that a pregnant woman might not have to eat anywhere near 400 pounds to endanger her unborn child's developing brain.[51] Pregnant women would do well to avoid fish entirely (even though they need an especially high intake of animal protein).

In the case of at least one American eating a lot of swordfish (contaminated at about the 1 ppm level) a diagnosis of mercury poisoning has definitely been given. Dr. Roger Herdman, deputy health commissioner for the state of New York, described for a U.S. Senate subcommittee the case of "Mrs. N.Y." At age forty-four, the mother of three children, Mrs. N.Y. went on a strict diet to lose weight. To avoid fish with a "fishy taste," Mrs. N.Y. started eating 10 ounces of swordfish per day. Within less than two years Mrs. N.Y. had lost plenty of weight (from 165 pounds to 120), but she had also developed a whole series of new symptoms: dizziness, losses of memory, quivering hands and tongue, difficulty in focusing her vision, trouble putting one foot in front of the other for proper walking. During 1966 Mrs. N.Y. sought help from the Neurological Institute at Columbia-Presbyterian Medical Center in New York City. There her case was diagnosed as "psychosomatic," and she was advised to try psychiatric therapy, which she did. After two and a half years of unsuccessful psychotherapy, Mrs. N.Y. was tested for mercury by the N.Y. state health department. "We believe hers to be the first case of human illness in this nation attributable to mercury poisoning from ordinary marketable food," said Dr. Herdman in reporting the case to the U.S. Senate. But, said Dr. Herdman, the illness of Mrs. N.Y. "cannot be regarded as merely an individual misfortune. There are bound to be others, if not many others, who are in the same boat."[52]

How many others are in the same boat? How many different groups of Americans—from blacks and whites in Alabama to Indians in Arizona and Eskimos in Alaska—supplement their diet with the cheapest form of animal protein? The number of people must be substantial. How many of these people will be damaged? It is impossible to say because the required controlled experiments cannot be carried out. And so instead of poisoning large human populations and sacrificing them to the microscope, scientists have taken measurements of a few people who were accidentally poisoned (in Japan and New Mexico) and then built mathematical models to simulate the status of poisons in the human body.

Two researchers at Berkeley have reported findings from this kind of mathematical modeling, which is carried out with the assistance of a computer. Dr. Robert Spear and Dr. Eddie Wei have calculated how much mercury would build up in different people's bodies if they ate so much fish for so long contaminated at a given level by mercury.[53] Spear and Wei conclude that allowing for normal statistical variations among individuals, out of a population eating one meal per week (170 grams or one-third of a pound) of fish contaminated at 1 ppm, after one year some 5 percent of the population would achieve a total body burden exceeding 6 milligrams of mercury per kilogram of body weight (mg/kg). If the fish were contaminated at 0.5 ppm and a person ate three meals per week (totaling 1 pound), the resulting body burden would be the same—6 mg/kg.

Spear and Wei point out that from empirical studies of Japanese people, it is known that death can result if body burdens reach 54 mg/kg. A kilogram (kg) is 2.2 pounds; a 154-pound person weighs 70 kg. Thus the 5 percent of the population achieving body burdens of 6 mg/kg would have less than a safety factor of 10 standing between themselves and possible death from mercury poisoning. Normally, public health authorities set safety standards which include safety factors of at least 100 to protect people from possible harm. A safety factor of less than 10 will not ordinarily be approved by public health officials. In addition, of course, safety factors are not supposed to protect people just from possible death; the purpose of public health standards is to protect against the least detectable physical harm. Beyond this, public health standards should also protect against the longest-delayed effects such as genetic mutations, birth defects, and cancer. According to officials of the U.S. Food and Drug Administration, these long-delayed effects have not been considered in setting the vast majority of the existing U.S. safety standards.[54]

The suppressed EPA report recommends tight safety standards in no uncertain terms. The report notes that the natural background level of mercury in uncontaminated freshwater fish is 0.2

ppm or less. The report then says that any body of fresh water containing a range of fish species with mercury levels above 0.2 ppm should be considered contaminated. Specifically, the EPA report says that public health authorities should investigate any such body of water to find the source of the contamination and stop it. *By far the largest part of the surface water system of the United States contains a range of fish species contaminated above 0.2 ppm.* No one has yet surveyed the entire United States satisfactorily for mercury pollution; when such a survey is completed, it may appear that all the surface waters of the United States have been contaminated to some extent.

The EPA makes clear what waters it defines as being hazardous:

> This Agency will consider that when a range of fish species in an inland or estuarine body of water have residue levels of mercury equal to or greater than 0.5 ppm (parts per million), that there is an indication of gross mercury pollution and that *consumption of fish from such a body of water would represent a hazard to health.*[55]

By this standard, something like half the surface waters of the United States would be classified as "grossly contaminated by mercury," and *eating fish from these waters would be strictly forbidden.* The American public has a right to be warned; further suppression of the EPA's view on the FDA's "safety" standard should certainly warrant Congressional attention. In the case of mercury, as in the case of lead, the federal Environmental Protection Agency has gone out of its way to keep important public health information from surfacing. Nevertheless, the American people have a clear right to know the facts.

Numerous reports in the popular press indicate that mercury has been found in old fish preserved in museums.[56] It is common now, in speaking in public about mercury pollution, to be told from the audience that mercury pollution has always been around and that, since it's nothing new, we should forget about it.

But mercury has been in use in instruments in scientific laboratories for well over 100 years. Since it is volatile at room temperatures, mercury is a consistent and pervasive contaminant of the air in most laboratories. In addition, because of its great germ-killing power, mercury has long been used as a preservative for biological samples. The mercury kills the microorganisms which would otherwise decompose any biological sample. Thus reports of mercury found in old fish are difficult to assess. Was mercury present as a contaminant in the original preservative? Was the sample carefully protected to prevent contamination by atmospheric mercury so common in laboratories?

However, we should assume that these old fish—forty years old or 100 years old or even 1,000 years old—*do* contain toxic levels of naturally-occurring mercury contamination. If this is the case, it simply means that people have gone for a long time without recognizing a potentially dangerous poison in their diet.

Most recently, press reports are appearing which indicate that many cats in Tokyo, Japan, have been found suffering from classic symptoms of mercury poisoning and with elevated levels of mercury detectable in their fur.[57] The cats—taken to veterinarians by their owners in an apparent epidemic poisoning—had been eating diets high in fish bought at public markets in Tokyo. So far no poisoning among humans in Tokyo has been linked to a fish diet—but this certainly doesn't mean damage isn't occurring. Japanese authorities have already warned their people not to eat large quantities of tuna.[58] Japanese people eat an average of 61 pounds of fish per person per year, whereas in the United States the average consumption is only 11 pounds per person per year. This is an important difference because, as J.G. Saha said in 1972, "It would then appear . . . that heavy fish eaters in Japan may be at risk."[59] And Dr. Saha was assuming that the 0.5 ppm safety standard would apply to the situation he was describing. Clearly, many scientists feel we have very little room for error here. Mercury contamination from whatever source, whether natural or not, must be treated with the greatest respect.

Here in the United States we have not yet recognized the severity of our mercury pollution problem, or its full extent, because it hasn't yet got as bad as it's going to get. Here is the most important point to emphasize: Even if we stop all mercury polluting today (which we're not doing), the problem will probably continue to get worse for perhaps 100 years or more. The buildup of mercury in lake bottoms and riverbeds will continue to methylate into the general biological environment. It seems likely to build up higher and higher in the bodies of freshwater (and possibly even some oceanic) fish. The President's Council on Environmental Quality revealed this unsettling information quietly in 1972, saying:

> There is evidence to indicate that the mercury problem may become even more acute and that levels in fresh water fish will rise over the coming years. There is a great amount of mercury deposited in bottom sediments, and the rate at which these deposits are absorbed by microorganisms will increase. This is because a larger number of microorganisms are being produced by the increase in amounts of nutrients entering U.S. waters from municipal sewage plants and from farmers' fertilized fields. Thus more mercury will be available to the fish which feed on the microorganisms, and in turn, to humans who feed on the fish.[60]

Dr. J. M. Wood, the well-known vitamin researcher from the University of Illinois, spoke in 1971 to a meeting of the Center for the Biology of Natural Systems in St. Louis.[61] Dr. Wood estimated that it might take 5,000 years before Lake St. Clair is rid of its deadly mercury deposits. (One Dow Chemical factory alone had been dumping more than 200 pounds of mercury per day directly into this lake for many years.)[62] Dr. Wood also reemphasized what others have pointed out: Trying to dredge out the mercury may result in increased mercury levels in the fish. Minamata Bay in Japan was dredged; the fish today are more toxic than they were before the dredging stirred the bottom sediments.

How bad will the mercury pollution problem ultimately get? "Twenty or thirty years from now," says Dr. Wood, "methyl mercury concentrations in most species of fish in inland waterways will be either lethal to fish or will render them unfit for human consumption."[63] Dr. Wood is saying that even if we stop polluting with mercury today, in twenty to thirty years all the fish in the United States will be dead or all the fish will be poisonous to human beings. Then, after that final delay, we may begin to understand what it is we're doing to the earth. But by that time it may well be too late.

> Whenever there is a long delay from the time of release of a pollutant to the time of its appearance in a harmful form, we know there will be an equally long delay from

the time of *control* of that pollutant to the time when its harmful effect finally decreases. In other words, any pollution control system based on instituting controls only when some harm is already detected will probably guarantee that the problem will get much worse before it gets better.[64]

—DONELLA MEADOWS and others (1972)

6

The Workplace Environment and Beyond

"I wonder how much of that dust gets into my system."

NO part of American life is more shrouded in secrecy than the workplace environment.[1] Here, where 84,200,000[2] American workers spend at least a quarter of their lives, almost no one has detailed facts on health conditions or environmental hazards. The only people who have been in a good position to gather facts—the owners and managers of the nation's workplaces—have pointedly ignored the problem for years. Unfortunately, many people calling themselves environmentalists have treated the problem with equal disdain, carefully defining "environment" to exclude the workplace. Now, however, at least the broad outlines of the situation have become clear: In extent, severity, and immediacy, the workplace is the No. 1 environmental health problem in America. The U.S. Surgeon General (then Dr. William Stewart) estimated not long ago that 65 percent of the American work force is exposed to toxic materials or harmful physical agents, and only 25 percent is adequately protected.[3] This should probably be viewed as an optimistic assessment of the situation.

For years the conventional wisdom has described the American workplace as an arena of continual progress; according to this view, work conditions are steadily getting a little better year after year. For a long time this view was valid. Now, however, the facts that are

112

available from government and industry sources leave little doubt that the workplace environment has been deteriorating for at least a decade and perhaps longer. Moreover, the nation's whole labor force has not undergone equal deterioration in working conditions. In the service sectors, for example, injury rates have declined during the past decade.[4] It is the 20,800,000 workers in the nation's manufacturing industries who have borne the brunt of the deterioration. We see this most dramatically in federal statistics on workplace injuries. There are two standard measures of workplace injuries: severity and frequency. Severity is a measure of the number of days lost because of disabling injuries per million employee hours worked; the federal Bureau of Labor Statistics considers this a measure of the seriousness of each injury. As Table 6-1 shows, for all U.S. manufacturing between 1958 and 1970 there has been a slight decrease in severity of injuries, averaging 0.3 percent.

The trend in frequency of injuries is much different. Frequency is a measure of the number of disabling injuries per million employee hours worked; the federal Bureau of Labor Statistics regards this as an indicator of the magnitude of the overall work injury problem. As an across-the-board average for all U.S. manufacturing, as Table 6-2 shows, the frequency of injuries increased 33.3 percent between 1958 and 1970.

For the average manufacturing worker in America the chances of having an accident on the job are 33.3 percent greater today than they were a decade ago (and the injury, when it happens, will be very nearly as severe as it was a decade ago).

Actually the available statistics badly underestimate the true situation. These statistics were developed under an injury-reporting system which purposely excluded millions of injuries from the official record. Under the old reporting system (which was changed somewhat in 1971), the federal Department of Labor was prohibited from making its own job injury count. The department had to rely on state governments to collect data, and the states were under no obligation to comply. Thus until 1971 there were only fourteen states even collecting data, and óf these, only six had reporting systems considered statistically adequate.[5] In addition, under the old system an "official" injury was one which caused a person to lose at least one full day of work beyond the day of injury; an injured person who received medical treatment and returned to work the same day or the following day was not officially counted as injured.

Table 6-1. Injury severity rates in selected U.S. industries, 1958 and 1970.

Industries	1958	1970	Change		13-year average
			Rate	Percent	
Mining:					
Coal mining and preparation	1 9,170	7,792	−1,378	−15.0	8,573
Metal mining and milling	2 4,147	3,238	−909	−21.9	3,586
Nonmetal mining and milling	2 2,244	2,624	380	16.9	2,514
Contract construction	2,496	2,100	−396	−15.8	2,307
Manufacturing	761	759	−2	−.3	719
Durable goods:					
Ordnance and accessories	198	284	86	43.4	283
Lumber and wood products	3,050	2,891	−159	−5.2	2,932
Furniture and fixtures	1,000	909	−91	−9.1	887
Stone, clay, and glass products	1,364	1,540	176	12.9	1,395
Primary metal industries	1,035	1,128	93	8.9	1,001
Fabricated metal products	1,023	1,003	−20	−1.9	922
Machinery, except electrical	525	583	58	11.0	573
Electrical equipment	282	333	51	18.1	282
Transportation equipment	479	488	9	1.8	447
Instruments and related products	261	270	9	3.4	274
Miscellaneous manufacturing	595	561	−34	−5.7	585
Nondurable goods:					
Food and kindred products	1,009	1,156	−147	−14.5	1,031
Tobacco manufactures	249	332	83	33.3	306
Textile mill products	550	579	29	5.3	514
Apparel and related products	231	207	−24	−10.3	166
Paper and allied products	993	937	−56	−5.6	875

Continued on following page

Industries	1958	1970	Change		13-year average
			Rate	Percent	
Nondurable goods:					
Printing and publishing	361	411	50	13.9	414
Chemicals and allied products	741	562	−179	−24.1	621
Petroleum and coal products	829	1,116	287	34.6	887
Rubber and plastics products	549	895	346	63.0	739
Leather and leather products	433	534	101	23.3	441
Transportation and public utilities:					
Local and interurban passenger transit	604	1,000	396	65.5	957
Motor freight transportation and warehousing	1,732	2,311	579	33.4	1,928
Communication	88	235	147	167.0	122
Electric, gas, and sanitary services	977	813	−164	−16.7	879
Wholesale and retail trade	[3] 553	452	−101	−18.2	482
Finance, insurance, and real estate:					
Banking	[3] 73	88	15	20.5	119
Insurance carriers	[3] 178	137	−41	−23.0	167
Real estate	[4] 763	341	−422	−55.3	504
Services:					
Personal services	475	276	−199	−41.8	430
Miscellaneous business services	563	310	−253	−44.9	499
Auto repair, services, and garages	1,005	427	−578	−57.5	789
Medical and other health services	264	264	0	0	251

[1] Rate not available for 1958; base year is 1960.
[2] Rate not available for 1958; base year is 1961.
[3] Rate not available for 1958; base year is 1962.
[4] Data available only for 1968 through 1970.

Table 6-1 adapted from **The President's Report on Occupational Safety and Health** (Washington, D.C.: U.S. Government Printing Office, May, 1972), p. 61.

Table 6-2. Injury frequency rates in selected U.S. industries, 1958 and 1970.

Industries	1958	1970	Change		13-year average
			Rate	Percent	
Mining:					
Coal mining and preparation	1 42.5	41.6	− 0.9	− 2.1	43.0
Metal mining and milling	2 23.6	23.7	.1	.4	23.0
Nonmetal mining and milling	3 20.6	24.1	3.5	17.0	21.9
Contract construction	30.9	28.0	− 2.9	− 9.3	29.1
Manufacturing	11.4	15.2	3.8	33.3	12.9
Durable goods:					
Ordnance and accessories	3.1	9.8	6.7	216.1	3.8
Lumber and wood products	37.0	34.1	− 2.9	−7.8	36.5
Furniture and fixtures	16.1	22.0	5.9	36.6	19.8
Stone, clay, and glass products	18.5	23.8	5.3	28.6	19.7
Primary metal industries	10.0	16.9	6.9	69.0	12.9
Fabricated metal products	14.5	22.4	7.9	54.5	18.0
Machinery, except electrical	9.8	14.0	4.2	42.8	11.9
Electrical equipment	4.9	8.1	3.2	65.3	6.1
Transportation equipment	6.0	7.9	1.9	31.6	6.7
Instruments and related products	5.5	7.9	2.4	43.6	6.6
Miscellaneous manufacturing	12.0	15.8	3.8	31.7	13.8
Nondurable goods:					
Food and kindred products	19.5	28.8	9.3	47.6	23.5
Tobacco manufactures	7.8	11.9	4.1	52.6	9.0
Textile mill products	9.0	10.4	1.4	15.6	9.7
Apparel and related products	6.0	7.7	1.7	28.3	6.8
Paper and allied products	11.4	13.9	2.5	21.9	13.1
Printing and publishing	8.8	11.7	2.9	32.9	10.1

Continued on following page

Industries	1958	1970	Change Rate	Change Percent	13-year average
Nondurable goods:					
Chemicals and allied products	7.5	8.5	1.0	13.3	7.7
Petroleum and coal products	6.7	11.3	4.6	68.6	8.4
Rubber and plastics products	8.7	18.6	9.9	113.7	12.8
Leather and leather products	10.9	15.2	4.3	39.5	13.3
Transportation and public utilities:					
Local and interurban passenger transit	13.3	23.9	10.6	79.7	17.9
Motor freight transportation and warehousing	28.9	35.5	6.4	22.1	31.8
Communication	.9	2.5	1.6	177.7	1.3
Electric, gas, and sanitary services	6.3	6.6	.3	4.7	5.9
Wholesale and retail trade	3 13.0	11.3	−1.7	−13.0	11.8
Finance, insurance, and real estate:					
Banking	2.3	2.4	.1	4.3	2.3
Insurance carriers	2.2	2.6	.4	18.2	2.3
Real estate	12.0	11.4	− .6	− 5.0	12.0
Services:					
Personal services	8.3	7.8	− .5	− 6.0	3.6
Miscellaneous business services	8.8	6.0	− 2.8	− 31.8	7.7
Auto repair, services, and garages	16.9	14.3	− 2.6	− 15.3	15.0
Medical and other health services	8.1	9.3	1.2	14.8	8.3

1 Rate not available for 1958; base year is 1960.
2 Rate not available for 1958; base year is 1961.
3 Rate not available for 1958; base year is 1962.

Table 6-2 adapted from **The President's Report on Occupational Safety and Health** (Washington, D.C.: U.S. Government Printing Office, May, 1972), p. 60.

For these reasons the official Labor Department count of job injuries in 1968 totaled fewer than 2,200,000; however, an independent check by the federal Department of Health, Education and Welfare revealed that 9,287,000 workers had been injured on the job in 1968. This represented more than 11 percent of the nation's total work force.[6] Today ten times as many worker days are lost from injuries as from strikes.[7]

The situation in work injuries is quite clear; in job-related illnesses the facts are not so well understood. The federal Department of Labor flatly declares that there are no data available on job-related illnesses;[8] however, the federal Department of Health, Education and Welfare estimates that 390,000 new cases of disabling occupational disease develop each year. HEW further estimates that 100,000 workers or ex-workers die each year from job-related diseases.[9] This is more than twice the number of people killed by automobile accidents each year.

Anthony Mazzocchi of the Oil, Chemical, Atomic Workers International Union (OCAW) said recently, "The people who run industrial America have a great stake in keeping us ignorant about what goes on in the workplace."[10] There is a good deal of evidence supporting this viewpoint. Even the Industrial Medical Association has pointed out that industrial managers and owners typically either fail to gather significant information about worker health or fail to make the data available to occupational health authorities.[11] There are reports of falsification of data by employers.[12] At a series of public meetings sponsored by the OCAW in 1969 and 1970 for the purpose of gathering information on work conditions in the chemical industry, several speakers stressed that companies which do perform tests on workers (chest X-rays or urinalyses, for example) routinely refuse to release the findings to the workers or their union representatives.[13] The companies seem to give a standard answer to these allegations: They say they can't release data because they're afraid of being sued (which leads us to believe that the findings may often be significant, if they could provoke suits).

As we have seen, for years federal law prohibited authorities from entering workplaces to gather data on environmental conditions. With the passage of the National Occupational Safety and Health Act of 1970 (also called the Williams-Steiger Act) officials now do have the right and the duty to inspect workplaces. At the end of 1971, after a year's total of 14,452 federal inspections,

35,839 citations had been issued for violations of health and safety regulations—an average of 2.48 violations cited on each inspection visit.[14] While most of the reported citations were for minor infractions, the remarkable number of violations is, in itself, an indication that the American workplace could definitely benefit from closer regulation and inspection.

Not all the health and safety violations in the workplace are minor, of course. At one of the OCAW conferences, for example, a worker named Harold Smith of the Wood Ridge Chemical Plant (Ventron Corporation, Wood Ridge, New Jersey), described work conditions on his job. Here, in a long quotation, is Mr. Smith's macabre humor and his wisdom:

> Most of our products are mercury. And we make a product called ROM, and we have three furnaces where we run a reactor of nitric acid and mercury into these three furnaces. Now these three furnaces have a tendency to explode. When they explode, they look like an atomic bomb. And when I first went to work in this department, I used to run in and out of this building all day long, trying to get away from these fumes.
>
> After we make this product, we turn it into a powder and we have to pack it. When we pack it, the dust from this here powder gets into our lungs, even though we have a mask on. And they're supposed to have a blower. But every time the doctor comes in to test me for the mercury that I inhale from this powder, they always tell me they should take me down to the still instead of sending scrap mercury, because I have more mercury in my system than we collect in the scrap container. . . .
>
> And it seems that when somebody comes around, the big bosses say, "Smith, don't you let that furnace explode while the big bosses are here." So then I have to keep that furnace going, and I say, "Listen, don't you explode." And I have three furnaces, and I have a name for each one of them—each one of the furnaces; I even talk to them. I say, "We got the big bosses in here; you better not explode." But it happens one day we had a New Jersey inspector in there and the furnace exploded. I was glad it exploded. And [the boss] he looked at me and said, "Why did you let that furnace explode?" I know that's what he was saying. But I was glad it exploded, because, you see if

these hazards don't show up when big wheels are there, they won't know about it. You see. So after, the big wheel saw what was happening. All these fumes went up in the air, just like an atomic bomb, and when it comes down, all this dust settles on the floor. And I might be over there packing, and I say to myself, "I wonder how much of that dust gets into my system."[15]

We have seen earlier that the long-range effects of mercury are devastating to the central nervous system; it seems reasonable to conclude that if conditions in Ventron Corporation's Wood Ridge plant do not change, workers there may age prematurely, losing their mental capacities before their time.

Workers have been subjected to severe occupational hazards since the earliest days of organized work. Metals in particular have long been a worker health problem; today metals continue to be a hazard, and they are now joined by thousands of synthetic organic chemicals.[16] Some of the problems are not life-or-death matters, yet they may cause a great deal of discomfort; for example, nickel, platinum, and chromium may cause skin eruptions.[17] Although skin disorders would not necessarily be classified as a "serious" health problem, anyone who has suffered from dermatitis (such as eczema) knows that it can be *extremely* discomforting. According to one estimate, 26 percent of the American work force is exposed to skin irritants; in workplaces employing fewer than 500 people, an estimated 36 percent of the workers are exposed to skin irritants.[18]

The most serious occupational diseases from exposure to chemicals occur after inhalation.[19] The lungs present to the atmosphere a very large surface covering about 55 square meters (M^2), or roughly the area represented by two tennis courts. (The tremendous complexity of the inner lung creates the large surface area.) The lungs are in the business of exchanging things—bringing oxygen into the bloodstream and getting carbon dioxide out. It is therefore no surprise that chemicals coming into contact with the lungs can often pass directly into the bloodstream; how a particular chemical behaves will depend on its physical and chemical state—for example, how large the particles are, and how soluble they may be, and their electrical charge.[20] Once a metal (or other foreign chemical) enters the bloodstream, the kidneys have to filter it out as best they can; for this reason, inhaled chemicals have a well-established record of attacking the kidneys and causing permanent renal damage.[21]

Lung diseases themselves may occur from numerous common materials—coal dust, for example, and cotton dust and asbestos. Nickel and chromium, cobalt and zinc have a long, well-established history of attacking the lungs of workers,[22] causing lung cancer and other serious diseases.

Another example would be arsenic. At least as early as 1960 the literature contained warnings about arsenic causing lung cancer.[23] Arsenic has been linked for some time now to cancers of the skin, lung, liver, mouth, esophagus, and larynx in people.[24] Yet it was early 1974 before federal health authorities finally came to grips with the situation. Workers at an Allied Chemical plant in Baltimore and at a Dow Chemical Company pesticide-manufacturing plant have lung cancer and lymphatic cancers six or seven times more frequently than the general public. Arsenic has been determined to be the cause, and as we write, the federal bureaucracy is once again gearing up to catch up and set standards, which it can then try to enforce.[25]

Other metals that bear watching in the workplace—because they shorten the lives of laboratory animals in controlled experiments—are gallium, germanium, cadmium, indium, tin, antimony, tellurium (a semimetal), mercury, and lead.[26]

A recent study by the National Institute of Occupational Safety and Health (NIOSH) shows that the incidence of lung cancer runs 229 percent higher in metal smelter workers than in nonsmelter workers. Heart disease rates and tuberculosis rates are elevated 18 percent and 41 percent over what would be expected in an average group of workers.[27] Metals are a serious airborne hazard inside the workplace—and, as we shall see, the dangers of the workplace environment now threaten large populations outside the workplace because pollution is expensive to contain.

As industrial chemistry becomes increasingly more complex, greater numbers of workers are exposed to new varieties of chemicals, most times without even being aware of the exposure.[28] For example, even today the word has not got around sufficiently to prevent cadmium poisonings among welders. In the welding of a cadmium-plated object (as when bolts are cut in a dismantling job), a yellowish fume is created. If the welder (or anyone nearby) inhales this fume, severe lung damage and even death can result. Craftspeople using cadmium-containing solders must beware.

Each year 300 to 500 new chemicals are introduced into the nation's industrial processes on a commercial scale.[29] As a general rule in the vast majority of cases, no testing is done to see whether these chemicals may cause long-term damage in workers exposed to them. Workers are thus used as experimental animals, the way guinea pigs and rats are used in the laboratory. Naturally, if toxic effects from a chemical can be observed, control measures will eventually be taken. But many of the most dangerous effects of chemicals result from long-term low-level exposure. Ill effects become obvious only after a long delay. In these cases (such as with beryllium and asbestos, which we will look at shortly), workers will be exposed to new substances for very long periods of time—twenty to forty years—before anyone recognizes a danger. By then, thousands or millions of people may have been exposed.

One often hears the argument nowadays that workers are protected by a full set of standards which establish "safe" levels for the chemicals in the workplace environment. However, examination of the situation makes it clear that workers are definitely not adequately protected. To begin with, safety standards exist for fewer than 6 percent of all the chemicals now in common use in U.S. industry. The federal goverment in 1971 listed 9,000 chemicals in large-scale commercial use, and health standards have been established for only about 450, or 5 percent, of these.[30] For the remaining 8,550 chemicals no standards exist.

Actually the 9,000 chemical figure badly understates the workplace that science has created, a workplace that Ray Davidson has characterized as "a chemical jungle full of lurking dangers."[31] Rachel Scott points out that the trade list *Chemical Sources* listed 17,000 industrial chemicals in 1958 and 41,000 different chemicals in 1970. Workers are potentially exposed to all of the 41,000 chemicals, while the general public's exposure is pretty much limited to the 9,000 commonest compounds[32] (except in the cases of "neighborhood disease," which we will discuss below).

Even in the cases where safety levels have been set, many of the standards themselves have been vigorously challenged on medical grounds. As Dr. Glenn Paulson of Rockefeller University has pointed out, the standards that *do* exist were established on a set of false assumptions. It was assumed that the typical worker is a healthy, young male who does not smoke cigarettes, does not breathe polluted urban air during his off-work hours, and who breathes only one chemical at a time in the workplace.[33] Obviously, for the vast majority of workers these assumptions are not valid. As the U.S. Public Health Service said not long ago, "The average worker spends only about one-fourth of his time at the workplace. The other three-fourths of his week is divided among a variety of environments, many of which are already hazardous or are rapidly tending in that direction."[34] On these grounds alone, workplace standards are based on assumptions which do not accurately reflect reality. In addition, at least half the American work force does not appear to be healthy; a survey in the early 1960s showed a little more than half of the nation's work force suffering from one or more chronic health conditions. Heart disease predominates.[35]

Obviously, multiple contaminants and multiple stresses are the rule rather than the exception on many jobs. For example, foundry workers are subjected to unhealthy levels of silica dust (fine sand) plus noise plus heat stress plus excessive carbon monoxide plus various resins,[36] in addition to metallic contaminants.

Another valid objection to existing standards is that they have been set without adequate knowledge of the primeval environments in which the human body evolved. For example, a team led by the well-known researcher Dr. Tsaihwa Chow has calculated that the present standard for lead in workplace air exceeds the amount of lead now found in "clean" (midocean) air by a factor of about 200,000. He has further calculated that, compared to the amount of lead found in air prior to industrialization (say 200 years ago), the existing workplace standard is higher by a factor of 20 million.[37] On the basis of these numbers, one might expect adverse health effects among workers at the present established "safe" level, which is 200 micrograms of lead per cubic meter of air ($200 \mu g/m^3$). This is, in fact, the case, as we have seen.[38] Serious lung changes have been recorded at $150 \mu g/m^3$. Anemia has been recorded among Japanese workers at lead levels of 140 to $150 \mu g/m^3$, and some health authorities suggest that the limit for lead should be $50 \mu g/m^3$ to protect worker health.[39] The lead industry continues to insist that 200 $\mu g/m^3$ is "safe" for workers,[40] but it is a fact that under existing conditions, lead workers suffer from a disproportionately high incidence of cerebrovascular disasters (strokes), which maim and kill.[41]

Another serious problem with present American workplace standards is that they do not take into account the existence of certain diseases. For example, in the case of the poisonous metal cadmium, British authorities have set their safety standard to prevent the development of chronic (delayed) cadmium disease. Americans, on the other hand, do not yet officially recognize the existence of chronic cadmium disease, and so the U.S. standard is set at a level intended to prevent acute symptoms only. Workers cannot feel well protected by such an irrational and arbitrary standard.[42]

Another instance of an inadequate safety standard is the one for asbestos. Like so many other toxic substances, asbestos had been in use for a long time before the full range of its effects on human health was even suspected. Only in the last four or five years has it been recognized as a possible trace metals problem. Asbestos is a rocklike mineral occurring in nature; it is the only mineral that occurs in long fibers which can actually be woven into cloth. Because it is fire-resistant, asbestos started coming into fireproofing uses around the turn of this century, and its uses have increased considerably ever since. Asbestos can now be found in thousands of consumer products—ironing board covers, pot holders, brake linings on cars and trucks, wallboard, plasterboard, floor tile, roofing, insulating, soundproofing, and on and on.[43] Two-thirds of all U.S. asbestos goes into building materials.

In 1890 the United States put 500 tons of asbestos into use, by 1920 the total had reached 200,000 tons, it reached 500,000 by 1930, and today we use about 4 million tons of asbestos each year.[44] Lung diseases from occupational exposure to asbestos were first reported by Greek and Roman physicians, and asbestos lung disease (called asbestosis) was observed in the modern world early this century.[45] Asbestos fibers break into tiny pieces during handling (as in weaving asbestos cloth) and become airborne, remaining in the atmosphere for very long periods of time. Workers and others breathe it in. In 1924 the first indisputable case of lung disease from asbestos appeared in medical literature, and in 1927 it was reported that people living in the neighborhood of an asbestos-processing plant had contracted asbestosis.[46] However, this first warning of so-called neighborhood disease was ignored for almost forty years. Why do we learn so slowly?

Table 6-3. Asbestos bodies in the lungs of a sample of persons in the general population at autopsy in New York City, 1966; preliminary analysis by sex and occupation.

Sex and occupation	Number of subjects	With asbestos bodies in lungs	
		Number	Percent
Men			
Blue collar	197	120	60
White collar	111	45	41
Women			
Housewives	47	14	30
Total	355	179	50

Table 6-3 adapted from Irving J. Selikoff and E. Guyler Hammond, "III. Community Effects of Nonoccupational Environmental Asbestos Exposure," **American Journal of Public Health**, Vol. LVIII (September, 1968), p. 1661.

Today there are an estimated 100,000 workers directly exposed to asbestos fibers on their jobs; however, an additional 3,500,000 construction workers—pipe fitters, carpenters, electricians, welders, plumbers, and others—are also exposed by the use of asbestos at construction sites.[47] Table 6–3 shows that blue-collar workers run the highest risk of asbestos damage.

Starting in 1935, it became recognized that asbestos fibers, in addition to causing the lung disease called asbestosis, cause two kinds of cancer. One, called mesothelioma, is cancer of the outer casing of the lungs or of the lining of the abdominal cavity. In addition to causing mesothelioma, asbestos also causes the familiar lung cancer. Mesothelioma and lung cancer are the biggest occupational health hazard from asbestos; half of all U.S. asbestos workers who die each year die of one form of cancer or another.[48] Yet the existing safety standard for asbestos in the workplace air is not set on the basis of preventing cancer but on the basis of preventing asbestosis. Mesothelioma and lung cancer are known to result from exposure levels below those which cause asbestosis. Therefore, the existing asbestos standard clearly cannot protect workers' health.[49]

In addition, the latest U.S. health standards[50] are set on the basis of counting large asbestos particles when, in fact, it is the very small particles which are most numerous and most dangerous and which enter the lungs most readily.[51] Dr. William Nicholson tells a story to illustrate the futility of measuring large asbestos particles when it's the small ones that are harmful. He says it's something like the man who was wandering around a lamp post when a policeman came up and asked him what he was doing. The man said, "I'm looking for a ten-dollar bill." The policeman didn't see a $10 bill, and he asked, "Where did you lose it?"

"I lost it down yonder over there," said the man, pointing.

"Then why are you looking over here?" asked the policeman.

"Because the light's better over here."[52]

The existing safety standard for asbestos has been established as a count of large asbestos particles (5 microns in diameter or larger) as seen under an ordinary optical microscope with magnification power of 400. However, as Dr. Irving J. Selikoff has pointed out, "a critical factor in studying asbestos air pollution is the utilization of techniques which will measure very small particles. It is unlikely that approaches which do not include the electron microscope [with magnification power of 40,000] will be effective."[53] Asbestos fibers smaller than 2 microns in diameter enter the lungs most readily. It seems certain that workers should not feel well protected by standards which do not take direct account of the smallest, most dangerous asbestos particles and which ignore the prevalence of cancers among asbestos workers.

Because of our experience with asbestos-induced "neighborhood disease," it has become clear in recent years that environmental health problems which were once thought to be "merely" occupational health problems must increasingly be viewed as threats to the general public. Take the case of the highly toxic metal beryllium. The dangerous properties of beryllium metal were first noted in medical literature in 1886. A second notice appeared in 1935. The first report of a U.S. injury from beryllium appeared during World War II, when this rare metal (lighter than aluminum, stronger than steel) began to come into military uses. The atomic energy industry started using larger and larger quantities of beryllium after the war, and by 1949 it was becoming widely recognized as a severe occupational health danger.[54] In 1952 a national case registry for beryllium injuries was first established, and more than 600 cases of beryllium disease have now been reported. Beryllium causes a lung disorder similar to emphysema. Emphysema is like a bad case of asthma or bronchitis which gets progressively worse until the patient can't walk down a hallway without feeling completely out of breath and exhausted. In advanced cases of emphysema, the patient cannnot blow out a candle held six inches in front of his or her mouth. It is a slow, debilitating, irreversible illness. Once you get it, you don't get rid of it. Beryllium causes a similar illness, just as debilitating as emphysema and distinguishable from it only by a specially-trained physician.

As with asbestos, cadmium, and lead, for a long time people thought that beryllium posed a hazard only to workers with regular occupational exposure. However, in the late 1950s, researchers began to discover neighborhood cases of beryllium disease. Neighborhood cases are caused by beryllium particles blowing around the neighborhood outside a factory which uses (or processes) beryllium. Actually the term "neighborhood" is misleading because it implies that the danger extends for only a block or two outside a factory. However, one of the sixty recorded "neighborhood" cases of beryllium disease occurred in a person who lived 5.3 miles from a factory which discharged small amounts of beryllium dust into the air. On the basis of this case, one might assume that there is some danger to people living inside a circle with an area of 88 square miles around a factory. Beryllium will probably never become a major environmental hazard either inside or outside the workplace because the world's supply of beryllium is limited; it's a very rare metal. However, it is already clear that beryllium is, and will continue to be, a deadly pollution problem of local industrial and neighborhood concern for the foreseeable future.[55]

A recent report has shown that beryllium reaches a large public through our national use of camping equipment. Mantel types of camp lanterns have beryllium metal in the mantel, and this beryllium enters the atmosphere immediately surrounding a lantern when it is lit. Lighting a mantel type of camp lantern inside an average tent or camping vehicle will produce a concentration of up to $18\mu g$ of beryllium per cubic meter of air ($18 \mu g/m^3$) for brief periods. The occupational standard for beryllium is $2\mu g/m^3$, but single exposures to concentrations as low as $25 \mu g/m^3$ have caused acute beryllium disease—so the danger from mantel types of camp lanterns is real.[56]

Asbestos is now widely recognized as a neighborhood pollutant of significant proportions. As noted above, the first cases of neighborhood disease from asbestos appeared in 1927. However, it was not until the neighborhood cases of beryllium disease came to light that officials really began looking hard for similar effects among people living near asbestos-using factories.[57] Then, in the period 1960 to 1963, three separate studies showed that asbestos could be found implanted in the lungs of at least 50 percent of all residents of big cities.[58] This came as a distinct shock to the medical community. From 1964 onward it was no longer possible to describe asbestos as "merely" an occupational health hazard to a "mere" 3,500,000 workers. Dr. Irving J. Selikoff has reviewed studies made in various cities, and he reports that asbestos fibers have been found in the lungs of 80 percent of the residents of London, England.[59] In New York City Selikoff's own studies have found asbestos bodies in the lungs of half of all residents; reporting these findings, Selikoff commented that if he had taken larger sections of lung to put under his microscope, he believes he would have found asbestos in the lungs of 100 percent of the residents of New York City.[60]

On the basis of these studies by Selikoff and others, the city of New York in February, 1972, banned the use of asbestos as a fireproofing material in new buildings.[61] Admirable and sensible as this ban may be, it certainly does not end the hazard from asbestos. For one thing, lung cancer and mesothelioma do not develop in people for twenty to fifty-five years after exposure to asbestos.[62] Thus, it will be well into the 1990s (or beyond) before we know whether we have *already* created a lung cancer epidemic from asbestos. In addition to this, the exposure of millions of average Americans has not ended. Asbestos fibers are extremely small, and they do not settle out of the air readily either with rainfall or by the force of gravity. The air of major cities may remain contaminated by asbestos for long periods of time after the use of asbestos as fireproofing has been banned. Auto brake linings and the demolition of buildings are a continuing source of asbestos fibers to the urban atmosphere. Thousands of do-it-yourself carpenters will continue to saw and sandpaper wallboard, soundproofing panels, and floor tile, never suspecting that they are exposing themselves (and their families) to invisible dangers from asbestos.

Since 1964, when it became obvious that asbestos was everywhere in America's urban areas, increasing attention has been focused on the precise mechanism whereby asbestos lung diseases occur. There are two competing theories to explain the mechanism whereby asbestos induces cancer. As we'll see, some researchers believe that the asbestos fibers contain trace metals (nickel and chromium) which act as enzyme poisons, inducing cancer by knocking out a protective mechanism in our lungs. On the other hand, there is now abundant evidence, which we have reviewed elsewhere,[63] showing that the physical size of the asbestos particles, and not their chemical nature, is what makes them dangerous. According to this second explanation, small persistent fibers thinner than 3 microns in diameter and longer than 20 microns in length cause cancer in rats whether those fibers are made of asbestos, fiber glass or aluminum.

Ninety-five percent of all commercial asbestos is a type called chrysotile.[64] Chrysotile is not a pure product; it is always contaminated, to one degree or another, with the metals nickel and chromium. It is now thought that asbestos causes cancer by introducing these toxic metals into the lung. The metals themselves seem to cause cancer indirectly. There is a natural family of compounds called benzpyrenes (BP for short) which have been suspected for a long time of causing cancer. However, the body has a natural mechanism for detoxifying BP. The detoxifying agent is an enzyme called BP-hydroxylase. It has now been discovered through laboratory experiments that the metals chromium and nickel reduce the functioning of BP-hydroxylase and thus prevent the normal destruction of BP in the body (lungs, liver, etc.) Thus it is hypothesized that when chromium and nickel are present in the lung, the body can no longer detoxify the carcinogenic agent BP, and cancer can develop.[65] This mechanism may help explain why there is a definitely synergistic (multiplier) effect between cigarette smoking and asbestos. Among people who are not asbestos workers but who smoke cigarettes, the chances of getting cancer are nine times the chances of the average nonsmoker getting cancer. Among asbestos workers who do not smoke cigarettes, the chances of dying of cancer are about two times greater than are the chances among the general population. However, among asbestos workers who smoke cigarettes, the chances of getting cancer are ninety times greater than are the chances of the average nonsmoker who doesn't work with asbestos. Thus, there is an obvious synergistic effect between asbestos and cigarettes.[66] Since cigarettes have been known for a long time to contain large quantities of the carcinogenic agent BP, it may be the nickel and chromium in asbestos which account for the synergistic effect. According to this view, the metals knock out the body's normal cancer-preventing mechanism at the same time cigarettes are putting a mass of cancer-causing agents directly into the lung. It is worth pointing out that chromium and nickel by themselves (not attached to asbestos fibers) have now been detected in the air of at least twenty American cities.[67] The source of these (other other) airborne metals is almost certainly industrial discharges into the outside environment.

Industrial discharges of toxic materials into the outside air cannot be viewed as an accident. In fact, according to a publication of the American Medical Association, modern industrialists, aware of the danger to workers from carcinogenic substances, have frequently adopted the practice of purposely dumping cancer-causing agents into the outside air.

> Many of the more recently constructed industrial plants are built in such a way that the machinery stands either free or in an open-air shelter, so that any injurious gases, fumes, or vapors may be readily dispersed into the surrounding atmosphere and thereby become harmless through dilution.
>
> While such a system may be suitable for some carcinogenic hazards, it seems to be unsuitable for others, especially for those which are due to agents that are not readily decomposed [such as metals] but accumulate on the ground and in the water and thus may gradually reach dangerous concentrations.[68]

Thus, although there is no denying that the 21 million industrial manufacturing workers of the United States are victims of the nation's largest, most critical, and least recognized environmental health problem, the hazardous environment they inhabit is, to some degree or another, shared by hundreds of thousands or millions of unsuspecting people who simply have the bad luck to live in a place polluted by industry. As Dr. Barry Commoner has said,

> To a significant extent the environmental crisis is an extension of problems that were once confined to the workplace to the community as a whole; likewise, the burden of these problems, which was once borne almost exclusively (and

still most heavily) by the worker, is now shared by the entire population.[69]

Exposing a person to a toxic chemical which shortens his life is tantamount to murder. . . . There's more crime in the workplace than there'll ever be in the streets, and no one is held accountable for it.[70]

—ANTHONY MAZZOCCHI (1970)

7

The Inner-City Environment

Chemical Warfare Against the Children of the Poor?

ONE hundred and thirty million Americans now live in cities or suburbs. For most of these people, manufacturing, transportation, and energy facilities (factories, cars, and power plants) have created a semipermanent pall over most of the sky.[1] Sometimes in Chicago as much as 40 percent of the sun's rays are filtered out by air pollutants before reaching the ground.[2] Here the people breathe a continually changing mix of industrial poisons: carbon monoxide, lead, cadmium, mercury, sulfur oxides, oxides of nitrogen, arsenic, chromium, nickel, asbestos, chlorinated hydrocarbons (like DDT, polychlorinated biphenyls, and toxaphene), plus more than 1,000 other pollutants. Perhaps not surprisingly, a high percentage of central city residents today are known to be suffering from multiple disease conditions. An extensive six-year study of the mental health of inner-city residents of New York in 1962 showed that 20 percent of the residents of midtown Manhattan could not be distinguished from inmates of mental institutions; one out of every five was judged insane. Four out of every 5 (80 percent) were judged "impaired" to one degree or another. Only 20 percent were found to be completely healthy.[3]

In addition to this severe indictment of modern urban life, there is much evidence that the inner-city poor endure physical as well as mental illnesses way out of proportion to their numbers. There are 8,000,000 inner-city poor (4,500,000 whites and 3,100,000 blacks, a figure determined partly because the government defines Mexican-Americans and Puerto Ricans as white). In this case, poverty is defined as the U.S. Department of Agriculture penuriously defined it in 1971: an income of $3,968 or less per year for each family of four persons (or $992 per person per year). It must be obvious that this income provides an extraordinarily spartan existence at best.

In addition to the 8 million inner-city poor falling below this official poverty line, there are 8 million others who live so close to poverty that their life chances, their opportunities, aren't statistically much improved over those of the official poor. Next to the endangered population of 20,800,000 workers in the nation's manufacturing industries, these 16,000,000 inner-city poor represent the nation's biggest immediate environmental health problem.[4] The conditions under which these people exist can only be called a grim travesty of any ecological or social ideal. The federal Environmental Protection Agency (EPA) recently described life for a typical inner-city resident:

> Coming home at night is not a relief for M.I. The city must be tolerated 24 hours a day. Some of his neighbors try to escape by deadening their awareness of the environment. M.I. feels such escape is not possible for him; he must try to stay well. He does not realize his life expectancy is seven years less than that of his suburban counterpart. . . . The world his kids grow up in is dirty, unhealthy and often unpleasant, but M.I. cannot afford to move elsewhere. His children are aware how the world treats their father. Their own experience in unpleasant surroundings, compounded by the poor quality of their schools and general discrimination, dim their outlook on the world and any prospects for a happy future.[5]

The average life span of the inner-city poor is seven years shorter than the life span of an upper-middle-income suburban counterpart. Among the inner-city poor, the chances of having a physical disability are four times as great as among suburban and rural people.

According to recent estimates prepared for the Environmental Protection Agency (then suppressed for more than a year after the findings were reported), at least half of all inner-city children never see a doctor; among these children, malnutrition is very common (especially iron and vitamin C deficiencies).

Among poor blacks (who represented 32 percent of all Americans living below the official poverty level in 1969), the chances of having hypertension (high blood pressure) are three times as great as the chances among affluent whites. In addition, disease rates are significantly higher among poor people (and blacks) for diabetes mellitus, kidney disease, gout, pneumonia, and cancer.[6]

For years the conventional wisdom has described the ghettos of America as places of steady improvement and progress. If the physical surroundings themselves were not improving, then at least the chances of escaping the ghetto (through so-called upward social mobility) were continually getting better. According to this view, everyone has always had a fair chance to escape, an equal opportunity for improvement. Now, however, the facts of modern urban life no longer support the conventional wisdom at all. In the past twenty-five years we have created a large class of permanently deprived inner-city victims who cannot escape the crippling circumstances of their birth. Some of these people, in turn, have changed the cities into frightening places for almost everyone.

Beyond the paralyzing direct effects of poverty itself, two kinds of bonds imprison the children of the inner-city poor: racial attitudes and environmental poisons. Generally speaking, the inner-city poor came to their present decayed urban circumstances only recently; because of this, they have inherited unique environmental problems, more insidious and more debilitating kinds of problems than past generations of ghetto dwellers had to face.[7] The recent history of today's inner-city poor begins with changes in agriculture following World War II. The slow industrialization of agriculture had been an obvious fact of American life since the 1830s (since the advent of McCormick reapers and subsequent inventions). However, in the southeastern United States the industrialization of agriculture lagged. After World War II events converged to make it profitable to industrialize Southeastern agriculture very rapidly. Black people emerged from World War II unable any longer to suppress their desire for fair and equal treatment in American society. They threatened the Southern status quo. But technology managed to save the old order for perhaps one more generation. Starting in the 1930s, chemists had created numerous commercially successful agricultural chemicals, such as the organochlorine compounds like DDT (which came into wide use as insecticides) and the phenoxy herbicides, such as the well-known weed killer 2,4,5-T. This new technology proved exceptionally profitable for petrochemical corporations. The new technology also made it possible to replace human labor with fossil fuel energy, plus large-scale machinery (metals plus energy combined), plus chemicals (inorganic fertilizers and synthetic organic herbicides and pesticides).

Recent changes in cotton agriculture, for example, were summarized in 1970 by the U.S. Department of Agriculture:

> The control panel of the planting-tractor is beginning to resemble an airliner's cockpit, for in addition to planting seed the operator is often controlling several operations . . . [There is] a rear-mounted planter equipped with fungicide, insecticide, and herbicide applicators. . . .
> Some of the greatest changes in cotton production technology have been in weed control. . . . It is now possible to virtually eliminate hand hoeing by using a combination of several herbicide sprays.
> . . . Machines harvested at least 96 per cent of our cotton crop in 1969 compared with 10 per cent 20 years ago. . . . Chemicals that defoliate the cotton plant of its leaves have helped make mechanical farming a success.[8]

Industrial farming entails huge outlays for equipment and for expertise. Small, inefficient farms have to sell out to the large landholders, many of which are now absentee-owner corporations. Southern landholders have, in the past generation, bulldozed their tenant-farmer shacks and packed the sharecroppers—black *and* white—off to live on welfare and food stamps in the decaying former homes of the well-to-do in Baltimore and Philadelphia and New York, in Detroit and Cleveland and Gary. A 1968 report from the federal government describes in a nutshell the trend we are discussing:

> While the proportion of Negroes to the total population has remained roughly the same since the turn of the century (between 10 and 12 per cent) a major distributional

transformation has taken place. Between 1940 and 1966 a net total of 3.7 million non-whites left the South for other regions of the United States.

The central cities of the metropolitan areas have absorbed most of the increase among Negroes since 1950—5.6 million of a total increase of 6.5 million Negroes. Fifty-six per cent of all Negroes now live in the central cities of metropolitan areas, while only about one-fourth of the white population of the United States now live in these cities. [9]

Once settled in the central city, the new residents have encountered and endured multiple stresses—lack of preparation for city life and the ever-present webs of racism that tie minority people to particular neighborhoods. As one medical researcher has put it, the new urban residents are "locked in closed, poisoned housing."[10]

The new residents have taken up the homes of the former well-to-do who began slowly leaving the cities as early as the 1920s (when the automobile first made possible the land-use patterns called suburbs). Since at least as far back as ancient Rome the well-to-do have painted their homes with expensive long-lasting lead-based paints. Although more expensive at first, lead-based paint lasts exceptionally long and has always therefore been considered a good investment by those who could afford the initial capital outlay. Now, however, these lead-painted homes are run-down, and the old lead paint, which is flaking and peeling, has become a severe and widespread poison threat to the children of the inner city.

Most houses in the central cities are not air-conditioned, so the windows are open a lot and the kids hang on the windowsills near the breeze. If they happen to chew the woodwork just a little—as some young children tend to do—they discover that the paint which is flaking off tastes good, sort of sweet and sour like lemon drops. This is ordinary white lead-based paint, and it is everywhere in America's central cities today. (For outdoor use, white lead is still a standard paint across the country.) In New York City in 1970, 58 percent of the apartments checked by the health department contained dangerous lead paint, and in Baltimore 60 percent.[11] As lead paint flakes off, it can contain as high as 66 percent pure metallic lead. A flake of lead smaller than a dime eaten once a day for longer than a month will probably bring on serious symptoms of lead poisoning,[12] perhaps including convulsions, extensive brain damage and death.

Twenty percent of all children between the ages of one year and five years develop a habit the medical profession calls pica. Pica means eating nonfood items; it is not an occasional tasting of a non-food item; it is a persistent and purposeful searching after nonfood substances and consuming them: paint, putty, plaster, dirt, ashes. It has been suggested that pica may develop in children who are iron-deficient; pica may be the growing body's attempt to take in essential iron to meet biochemical physical needs.[13] Iron deficiency is now known to be extremely widespread among the children of poor families (and even among the children of the well-to-do).[14] Further, it has recently been reported that lead has especially severe toxic effects upon children who are deficient in iron, protein, or calcium.[15]

Ordinarily pica is controlled by the person watching over a child; mothers or baby-sitters usually prevent children from eating significant quantities of nonfood items. But in a family under stress, malnourished, on welfare, unable to find work (much less, meaningful work), crowded into polluted slums, pica can go uncontrolled long enough to bring on lead poisoning.[16] Kids can snitch paint off a peeling fence in a back alley or lie under a porch and pick it off the bottom of the railing. Taste treats are rare enough even in the well-to-do world; to a slum child lead paint can be a delicacy. "It's surprisingly good," says Dr. Hyman Merenstein at Kings County Hospital, Brooklyn, N.Y. Dr. Merenstein couldn't understand why children ate lead paint, so he tried some himself. "It has a sweet taste, kind of like a cordial candy—with a kind of alcoholic after-taste," he says.[17]

Recent measurements show that there is enough lead on the floors and on the playground dirt in slum areas to poison infant children—crawlers and toddlers. Children who get lead on their hands and then suck their fingers or lick their hand can get sick.[18] And it isn't by any means all flaking off the walls. A lot of it is falling out of the air (put there by automobile tailpipes).

After one to three months of eating paint (or plaster or putty, which also usually contain lead), a child may begin to show one or more signs of lead poisoning: loss of appetite, constipation, weakness, awkwardness or clumsiness, fatigue, irritability, lethargy, headache, abdominal pain, sometimes vomiting.[19] Recent evidence indicates lead can cause hyperactivity in children.[20] In children, lead has a special affinity for the central nervous system; the ages from birth to two are a critical time in the growth of the human brain and associated nerves. Lead destroys brain tissue and the related systems of nerves leading outward from the spinal column which control the entire body.[21]

If lead poisoning is caught during its early stages—and *if the child is never again exposed to lead*—there is an even chance or better that no long-term damage will result. However, most cases of lead poisoning are not caught early, and most of the cases that *are* caught get sent home in a few days, and home is the poisoned place. This is an important point. It has now been established by physicians that the symptoms of lead poisoning are worse the second time and worse again the third time—the effect of repeated poisoning is cumulative. One team of doctors wrote in 1966:

> Re-exposure and continuing exposure to lead increases mortality [death] and morbidity [illness]. As a result of re-exposure, the subclinical [symptom-free] case may begin to develop gastro-intestinal and then neurological symptoms. The mild case may develop ataxia [staggering, inability to control arms and legs], then seizures, and finally, a fulminant encephalopathy [devastating pressure on the brain].[22]

Because reexposure to lead is certain to worsen a child's chances of recovering, medical authorities now agree that once a child has been diagnosed as a lead-poisoning case, he or she should never be returned to the home where the poisoning occurred unless the source of poison has been chemically identified and removed. One researcher says it is "axiomatic" that a child should never be returned to the poisoned environment.[23] Dr. J. Julian Chishold Jr., who is probably the nation's best-known expert on childhood lead poisoning, says, "Once a child has suffered a serious metabolic insult such as lead poisoning, he should never be returned to a lead-contaminated house." [24] However, even in a city which is aware of the lead-poisoning problem among children, it takes from six to eleven weeks for health authorities to see that a home has been made safe.[25] Making an old home poison-free isn't easy. Simply painting over old lead doesn't help; peeling exposes the oldest lay-ers. Removing lead paint is itself hazardous (the dust can be lethal) and expensive, so usually health authorities settle for four-foot-high wooden wall coverings which make it impossible for children to eat paint off the lower parts of walls. (The back fence or the porch or someplace else outdoors in the neighborhood can continue to supply toxic paint to a pica child.) In the meantime, while the six to eleven weeks go by (with the health department, the landlord, the contractor discussing and debating), the poisoned child spends most of that time back in the environment that poisoned her or him to begin with. No one has published studies, so far as we know, showing what percentage of lead-poisoned children are treated more than once for symptoms. The record does show that a remarkably small number of children actually receive *any* treatment. In New York City, where an estimated 122,000 children are considered "at risk," fewer than 3,000 were actually given medical treatment during 1970.[26] The remainder of the "at risk" population got along the best they could without medical help. Are we poisoning up to 10 or 20 percent or more of the children of the poor each year? The record seems to indicate that we are.[27]

Blood samples taken at random have shown repeatedly that a high percentage of slum children have elevated levels of lead in their blood. In Washington, D.C., where all two-year-olds coming to clinics (for whatever reason) automatically receive a blood test for lead poisoning, 23 percent have blood-lead levels above 40 micrograms per 100 milliliters (μg/100ml); 14 percent have blood lead above the 50 μg level.[28] In Chicago, 8.5 percent of a large group of slum children had blood lead above 50 μg.[29] Fifty micrograms is definitely within the effective poison-damage range; many physicians feel that poisoning can definitely begin at 40 μg, and some hold that 25 μg is enough to cause suspicion of poisoning in children.[30] No one knows how much it takes to harm a fetus during critical periods of brain and nerve development. It has been reported from several cities that adults living in the inner city now have something under 30 μg lead per 100 ml of blood, but in some urban places children now *average* between 40 and 50 μg per 100 ml.[31] If this is fully confirmed, our own statement, that we are subtly poisoning 10 to 20 percent of the nation's nonwhite children, may be quite low.[32]

Without redistributing income and wealth to eliminate poverty, the problem of lead poisoning of children in our slums seems to be

insoluble. "You're talking about tearing down an old city and building a new one," says one Washington, D.C., official.[33] An official in
St. Louis, Missouri, has estimated that it would cost $100,000,000
to remove the lead-paint hazard from the slums just in St. Louis.
Nationally, the size of the job is staggering.[34] Clearly the private
sector of the economy has no incentives to attempt solutions, and
characteristically, the federal government has failed to attempt solutions which are adequate to the magnitude of the problem. After
making headlines for three years or better, Congress finally passed
a law which appropriated $30,000,000 to solve the national lead-
paint problem. But the Nixon-Ford administration has released
only a small part of the full appropriation.[35] In processing its 2,649
known lead-poisoning cases New York City alone spent $2,400,000
in 1970[36]—so we can see clearly that without funding at the level of
hundreds of millions of dollars, the problem of lead poisoning is
going to get worse and not better.

Lead poisoning is the kind of disease that makes itself known
only if health authorities aggressively look for it. A federal expenditure of a few millions of dollars constitutes "not looking." It
amounts to a racist federal policy, a policy of silent violence against
the children of the poor. In New York City in 1950 there was only 1
reported case of lead poisoning; by 1969 health authorities were
really gearing up to look for the disease and they found more than
700 cases.[37] By 1970 mass blood testing had got started, and 2,649
cases emerged.[38] After seeing how many cases show up among a
population when mass blood testing begins, federal officials estimate that up to 400,000 American children now suffer from some
degree of lead poisoning.[39]

Lead builds up in the body over time, about 91 percent of it being stored in the bones. However, numerous illnesses can cause
stored bone lead to reenter the bloodstream at a later date and
bring on symptoms of lead poisoning.[40] In one large group of Aus-
tralian children poisoned by water collected off a lead roof while
they were very young, serious poisoning symptoms only showed up
as massive kidney damage and cardiovascular disease twenty-five
years later.[41] Any illness that results in a high fever can remobilize
bone-stored lead, as can cortisone medication (now used widely in
treating arthritis).[42] Also any infection that severely changes the
body's acid balance (bringing on a condition called acidosis) can

mobilize lead stored in bones. Perhaps most important, plain malnutrition can cause the body to remobilize stored lead.[43]

"The number of cases of lead poisoning depends upon how hard people look," says Dr. Evan Charney, associate professor of pediatrics at the University of Rochester School of Medicine and Dentistry. "If you live in an American city with a slum population and you don't have many cases of lead poisoning, your health department isn't doing its job," Dr. Charney says flatly.[44] Even when physicians and public health authorities *do* look for the disease, it can prove very elusive and hard to diagnose. Frequently lead poisoning has been diagnosed as influenza or as a digestive disorder. Typically, the physician prescribes an antibiotic and sends the child home.[45] As one doctor recently described it, the symptoms of lead poisoning are "more diffuse and difficult to interpret [in children] compared to adults."[46] In severe cases of poisoning, it is especially critical to make a correct diagnosis. The only way to diagnose advanced lead poisoning with certainty is by a spinal tap, and the spinal tap itself can be dangerous.[47] In fact, most therapy against lead poisoning can be dangerous. The commonest way to treat a lead-poisoned child is to administer a series of painful injections of a chelating agent. We recall that chelating agents bind metals into tight chemical structures and carry the bound metals out of the body. There are three common chelating agents used in lead therapy; they go by the names of EDTA, BAL, and penicillamine. Unfortunately these agents themselves—especially EDTA—may, under certain circumstances, severely damage a person's kidneys, so it is desirable never to administer chelating agents unless absolutely necessary.[48] Thus, a correct diagnosis of lead poisoning becomes important from the patient's viewpoint. (We are born with a single set of kidneys, and if we wear those out, we die.)

Among those children who are known to have experienced moderate to severe lead poisoning, 22 percent show permanent mental retardation—with IQs as low as 20—indicating massive damage to the central nervous system. This damage may show up in other ways besides mental retardation: recurrent epileptic seizures, cerebral palsy, blindness, and damage to the kidneys, heart, and muscular systems.[49] At least one researcher, Dr. Joseph Dimino, has concluded that subclinical lead poisoning (poisoning in which no symptoms become obvious right away, but in which lead can be measured in the blood above known safe levels) is responsible for widespread underachievement among children. Dr. Dimino concludes that children would do much better on mental tests, that they would be happier, and that they would feel better if they took less lead into their bodies.[50] As we saw in an earlier chapter, there is also evidence that they might live longer and suffer from fewer infections. Dr. Henry Schroeder and Dr. Harriet L. Hardy have suggested that we would *all* feel better if we took less lead into our bodies.

Recent reports of zoo animals poisoned by lead—some of it in paint on their cages, but most of it suspected of being airborne—have reemphasized the plight of the inner-city poor. Zoo animals in the Staten Island and Bronx zoos in New York and at the municipal zoo in Detroit have been reported dying in excessive numbers from lead poisoning: snakes, owls, mice, and monkeys, among others, have been lethally poisoned. Reporting the zoo story in *Science,* Robert J. Bazell noted correctly: "The findings have ominous implications for the people who live in that area of the city." [51]

> The number of cases of lead poisoning depends upon how hard people look. If you live in an American city with a slum population and you don't have many cases of lead poisoning, your health department isn't doing its job.[52]
>
> —Dr. Evan Charney (1970)

8

"Ill Fares the Land, to Hastening Ills a Prey"[1]

Industrial Agriculture

IT has been obvious for more than a decade that America's cities are sick.[2] However, the less obvious illnesses of the rural countryside will finally prove to be more important. From the viewpoint of amenities and services and quality of life, the cities have been in deepening trouble for several decades; the rural areas, however, appear to be flirting with a more far-reaching kind of agricultural/ ecological collapse, threatening the nation's long-term food supply. As an international symposium of ecologists recently concluded,

> When the practice of industrial agriculture is interpreted using current knowledge of ecosystems, a picture emerges which suggests that the future dependability of one of our most essential biological systems is in grave doubt.[3]

The group went on to note that although the practice of industrial agriculture has been eminently successful for more than thirty years, "There is, however, reason to suspect that major difficulties will develop in the next few decades."[4]

One of the main features of industrial agriculture is the starkly simplified ecosystem—for example, mile after mile of fields planted to a single stand of grass (such as corn). Simple ecosystems make the use of big farm machinery economical. However, the creation

143

of large, simple ecosystems has adverse side effects which have gone unrecognized for too long. Simple ecosystems are subject to devastating attack by insect pests or invasion by weeds or by plant-disease organisms. Without constant management, constant inputs of chemicals and fossil fuel energy, modern agricultural ecosystems quickly collapse. Major outbreaks of pests have been recorded in American agriculture since the early 1870s—just 100 years ago—and chemical pest controls have been employed regularly since the 1880s and 1890s.[5]

The first chemical pesticides were compounds of the poisonous metals arsenic, lead and copper. (They poison enzymes in bugs just as they do in people.) After World War I, methyl mercury compounds—the most poisonous of all—began to come into use. Today even more exotic metals have found their way into pesticidal use: the highly toxic metal thallium, for example, and cadmium, tellurium, and selenium.[6] Many pesticidal compounds contain a combination of poisonous metals. One recent listing shows 141 different metallic compounds contained in 112 pesticide formulations.[7]

The use of metals in pesticides has been declining in recent years because it has now been widely recognized among agronomists that metals don't break down in the environment. The metallic compounds are thus classified as permanent pesticides.[8] They can remain biologically active in fields and orchards for twenty to fifty years and in some cases even longer. For example, H. J. M. Bowen wrote in 1966:

> The use of [the toxic metal] arsenic as an herbicide and insecticide is decreasing in view of its toxic and carcinogenic effects on mammals. Until recent years about 3×10^7 kg/year of arsenic was applied to crops. . . . As a result, the arsenic content of arable fields in some parts of the world has risen so high that crops taken from them will continue to contain excessive arsenic for many years.[9]

Since World War II the cheaper, less permanent organochlorine pesticides (DDT and related compounds) have come into wide use, replacing the metallic compounds. Since the organochlorine compounds can remain active in the environment for two to fifteen years or longer, they are classified as persistent pesticides, and they have now become recognized by biologists and biochemists as a worldwide danger to many life forms and many ecological systems.[10]

All chemical pesticides lead to simplification of ecosystems. Because most pesticides are broad-spectrum poisons, they not only kill the target species but also wipe out many of the predators and parasites nature provides to keep each species in check. Most insect species in natural ecosystems have something like 100 predatory and parasitic enemies; the tremendous complexity and interrelatedness of these predator-prey-parasite relationships helps make ecosystems stable.[11] Complexity in itself is important to maintain because it lends stability to the earth's entire capacity to support life. By wiping out large numbers of predatory non-pest species, thus simplifying ecosystems, chemical pesticides favor the development of unstable, unbalanced ecosystems which require constant management lest they collapse.

For a time, chemicals may keep an important pest under control. However, insects and microorganisms fairly quickly develop resistance to chemical poisons. They do this by natural genetic (Darwinian) evolution, adapting to the presence of new environmental conditions and stresses. An application of a poison kills all but the hardiest; but it isn't long before those few hardy survivors have regenerated their numbers. Now they are resistant to the poison, and a new toxic agent must be developed to kill them. But some survive the new toxic agent . . . and so resistance is ultimately an insoluble problem. Since World War II more than 200 different insect pests have developed resistance to one or more pesticidal formulations.[12] In addition, by a phenomenon known as cross-resistance, pests adapting to the presence of one poison frequently show definite resistance to other chemically similar synthetic poisons. So today's pest and non-pest species are both to some degree developing resistance to tomorrow's chemicals. When today's non-pest emerges as tomorrow's pest, it will already be resistant to part of the farmer's chemical artillery.[13]

By these various means, it is possible for ecosystems to become dependent on pesticides. Thus it becomes possible to imagine large portions of the biosphere dependent on human injections of poisonous chemicals to maintain stability and balance (after our blind meddling has ruined the ability of ecosystems to maintain their own equilibrium). An international group of environmental specialists, gathered at the Massachusetts Institute of Technology in 1970, wrote:

> For many of our crops on which pesticide use is heavy, the number of pests requiring control increases through time. . . . Thus pesticides not only create the demand for future use (addiction), but they also create the demand to use more pesticides more often (habituation). Our agricultural system is already heavily locked into this process, and it is now spreading to the developing countries. It is also spreading into forest management. Pesticides are becoming increasingly "necessary" in more and more places.
>
> Before the entire biosphere is "hooked" on pesticides, an alternative means of coping with pests should be developed.[14]

The forces that introduced chemical pesticides into American agriculture are the same forces that accelerated our dependence on inorganic chemical fertilizers and large-scale farm machinery. All these forces have their roots at least as deep as the early nineteenth century, but they have grown most remarkably just in the last thirty years.

Beginning in the period around World War I— 1910–1920—it became clear that the most basic problem facing American farmers was the problem of abundance. The American soil was too rich, the American farmer too skillful. Given a commitment to a market economy, an abundance of agricultural products guaranteed that the price of farm products would fall. The falling prices of farm products throughout the 1920s kept the American farmer in continual economic depression. The sudden visibility of the Great Depression in 1929 merely made things worse; by 1933 American farmers were receiving 56 percent less for their produce than they had received a decade before.[15]

Modifying measures that had first been adopted on an emergency basis during World War I, the Roosevelt New Dealers from 1933 to 1938 passed laws forming the ground rules of today's farm subsidy or production control programs. Although the farm subsidy program is rather complex at the day-to-day working level, in principle it is simple enough: By various means the government pays farmers to take part of their land out of use each year, to restrict their acreage.

This production control program can more or less solve the problem of agricultural surpluses, and it *does* solve the problem without destroying all appearances of a commitment to a market economy.[16] However, the acreage-restriction program has had important destructive side effects which the nation has only recently begun to recognize. From its earliest beginnings better than fifty years ago the acreage-control program has especially rewarded those farmers who could restrict their acreage but still increase their per acre yield each year. The key to success within the farm subsidy program lay in increasing per acre crop yields. And so the pattern of the past fifty years has been a steadily accelerating dependence on tremendous injections of energy and metals (fossil fuels and farm machinery) plus inorganic fertilizers plus pesticides plus water, to maximize per acre yields. The effect of the acreage-control programs has been to subsidize and promote modes of farming which require enormous amounts of capital. For example, investment in farm machinery has increased from $3 billion to more than $30 billion since World War II.[17] To compete in the new field of industrial (capital-intensive) farming, the farmer needs substantial financing. The U.S. Department of Agriculture now estimates that an investment of $100,000 is the minimum required to begin efficient farming. Further, the department estimates that an investment of $200,000 is the minimum required to provide sufficient farm income to feed, house, and clothe a family of four and provide a college education for two children.[18]

As a result of these trends, the small farmer is driven off the land. In 1935 there were 6,800,000 American farms; today there are fewer than 2,900,000. Each year between 80,000 and 100,000 independent farm units are gobbled up by the large farm producers.[19] As the number of farms has dropped, the number of people living on farms has naturally dropped too. Since 1945 farm population has decreased from 24,400,000 to 10,300,000, from 17.5 percent of the total national population to 5.1 percent.[20] Of the 2,900,000 farms remaining in America today, approximately 1,000,000 have gross sales of $10,000 or more. These farms produce 85 percent of our food and fiber. That leaves approximately 1,900,000 farms producing only 15 percent of America's food and fiber. These farmers have incomes below $10,000; about 1,300,000 of these farmers, in fact, have incomes below $2,500 a year.

At the other end of the scale, the top end, we have a handful of successful farmers—200,000 of them. They each gross an average $40,000-plus annually, and their average net return is $24,000 annually; between them they produce half of America's food and fiber. At the very top of the scale we find the farmer of the future: the agribusiness corporation. In 1965 (the latest year for which we have data) there were 17,578 corporations engaged in farming but this handful of giants already accounted for some 10 percent of all agricultural production.[21]

The biggest American farming corporation in 1964 had gross receipts of $432 million; this was more than 1 percent of the total gross receipts of all farmers in the United States combined. One hundred corporations like this one could equal the growing capacity of all the 2,900,000 American farms which exist now. Dr. Vernon Ruttan, chairman of the Department of Agricultural Economics at the University of Minnesota, views the rapid trend toward agribusiness modes and concludes, "A food and fiber industry in which 80–90 percent of farm output could be produced by 50,000 to 100,000 production units is not only technically feasible but is in the process of evolving."[22]

The trend toward largeness in agriculture, the trend toward capital intensive farming operations has forced radical changes in American farming practices. The Department of Agriculture's 1970 Yearbook described a typical post-World War II change in crop production:

> Up to World War II, corn typically was grown in a 3-year rotation of corn-oats-clover, without fertilizer . . . and the Corn Belt yield was about 38 bushels an acre.
> Today, corn is seldom rotated. Leading growers typically fertilize with 150 pounds of nitrogen . . . and get yields of 130 to 150 bushels an acre. The average Corn Belt yield is 90 to 100 bushels.[23]

The trend toward higher per acre yields naturally drives the soil harder and harder, removing nutrients from the soil at accelerating rates. To compensate for this, in the last twenty-five years we have rapidly increased our use of inorganic chemical fertilizers (chiefly nitrogen and phosphorus) to restore the nutrient values of the soil. (See Figure 8-1.) This practice itself has several important destructive side effects. First, up to 25 percent of the fertilizer put onto a field doesn't stay there but runs off with rain or irrigation water into the nearest body of water, where it causes several problems. Most obviously, it stimulates plant growth.

The stimulation of plant growth in a lake or stream shortens the productive life of that body of water in a process called excessive eutrophication. Secondly, phosphate fertilizer contributes poisonous cadmium and lead and other toxic trace elements to soil and plants and to plant consumers.

Thirdly, increasing reliance on inorganic fertilizers may change the nutritional quality of the crops grown. There is a lot of evidence to support this view, as we shall see in the next chapter. Heavy fertilization can lead to nutritional imbalance in soils unless elaborate soil analysis and treatment are routinely carried out.

There is one last important point about the dependence of industrial agriculture on inorganic fertilizers. The current practice of industrial agriculture seems to be rapidly depleting the world's known and inferred reserves of essential phosphorus. The prestigious Institute of Ecology recently estimated that our accelerating use of phosphate rock and superphosphate will exhaust the world's entire reserves of phosphorus just sixty to ninety years from now.[24] This is the grimmest prediction we have encountered. The institute elaborated on the magnitude of the problem that is suggested by the loss of phosphorus reserves. At current rates of population growth (1.9 percent to 2.1 percent per year worldwide) and at current rates of growth of phosphorus fertilizer use (5.25 percent increase each year for a doubling time of 13.3 years), world phosphate rock deposits may run out at a time when world population has grown to perhaps 11 billion. At present world population growth rates, we will reach 11 billion in fifty-two years, or in the year 2028. At that time, says the institute, the world's agricultural operations would have to revert to modes of production which probably cannot support more than 1 to 2 billion people around the globe. According to the institute, this could leave up to 10 billion people literally without means of subsistence, reducing world population to 10 or 20 percent of its former levels in a drastically short time. Such unprecedented destruction of human life must be left to the reader's imagination to comprehend.[25]

150

Figure 8-1 **WORLD FERTILIZER CONSUMPTION**

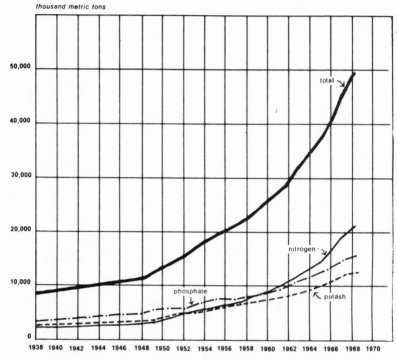

thousand metric tons

World fertilizer consumption is increasing exponentially, with a doubling time of about 10 years. Total use is now five times greater than it was during World War II.

NOTE: Figures do not include the USSR or the People's Republic of China.

SOURCES: UN Department of Economic and Social Affairs, *Statistical Yearbook 1955, Statistical Yearbook 1960,* and *Statistical Yearbook 1970* (New York: United Nations, 1956, 1961, and 1971).

When the practice of industrial agriculture is interpreted using current knowledge of ecosystems, a picture emerges which suggests that the future dependability of such agriculture is in grave doubt.[26]

—THE INSTITUTE OF ECOLOGY (1972)

9

The Deficiency State

Metals (and Other Essential Nutrients) Missing from the American Diet

AN estimated 116 million Americans (56 percent of the country's total population) now eat food containing inadequate amounts of the essential metal calcium; 57 percent (117 million people) eat a daily diet lacking the Recommended Daily Allowance (RDA)* of the essential metal iron.[1] In regard to vitamin intake, the story is nearly the same: 31 percent of all Americans eat food containing inadequate vitamin A; 57 percent of all Americans fail to take in the RDA for vitamin C. Among the B vitamins it's the same: 56 percent of the people get inadequate thiamine (B_1); 43 percent get inadequate niacin (nicotinic acid); 40 percent get inadequate riboflavin (B_2).

Moreover, the adequacy of the RDA standards themselves has been challenged because they are average numbers attempting to suggest nutritional limits for average people. Naturally-occurring genetic variations among people are known to cause wide differences in nutrient requirements among individuals. Therefore, many individuals may be deficient in one or more nutrients despite dietary intakes adequate by RDA standards. There does seem to be a good deal of evidence supporting this view.[2]

*The RDA is set by the National Academy of Sciences; see *Recommended Dietary Allowances* [Eighth Revised Edition], Washington, D.C.: National Academy of Sciences, 1974.

Even if we accept the RDA values as useful guidelines, the picture of malnutrition in the United States today appears to be of large proportions. It is not just poor people who eat an inadequate diet today; malnutrition seems to be common among the well-to-do (though it does remain true, in general, that the poor eat a much worse diet than the rich). Malnutrition among the rich stems from ignorance and inadequate parental guidance in youngsters' diets.

A team of analysts recently surveyed all nutrition studies which were completed in the United States between 1950 and 1967. Reviewing all the evidence, they concluded, "We believe that the studies which have been done in the past two decades have indicated that, to one degree or another, there are nutritional problems in the United States affecting virtually all age groups and segments of the population."[3] Half a dozen separate studies in the past two decades have shown, for example, that the highest incidence of infant malnutrition apparently occurs among the well-to-do. Babies born to the wives of college professors and successful businessmen evidently start life somewhat malnourished in the large majority of American cases.

There is now considerable evidence indicating that malnutrition is worsening in this country. For example, the federal Department of Agriculture measured the nutrient content of family diets in 1955 and again in 1965. In 1955 the USDA found "good" diets among 60 percent of all families; by 1965 the number of "good" diets had fallen to 50 percent. In 1955 the USDA found "poor" diets among 15 percent of the nation's families; by 1965 the number of "poor" diets had risen to 20 percent.[4]

The researchers who reviewed two decades of nutrition studies tried to explain why malnutrition might be worsening in the United States. They concluded that American nutrition is getting worse because of technological changes which took place after World War II. At that time "significant changes in food supply, enrichment, and fortification of foods [and] methods of preservation of food . . . occurred."[5]

It is indisputable that food processing has altered the vitamin and mineral content of foods. Food processing has also made it possible to cloak calories (energy) in taste treats that contain zero vitamins and minerals, thus giving rise to a class of foods now known as empty calories.[6]

However, the problem of malnutrition strikes deeper into American habits of thought than mere technological changes in food processing. Basic agricultural practices and basic habits of thought among medical doctors have also contributed heavily to the current worsening wave of malnutrition. We'll look briefly at medical habits of thought first. The analysts who surveyed nutrition studies, 1950–1967, made this important point: "Unfortunately, in this country interest in evaluating nutritional status has not run high among biochemists and has been almost completely lacking among physicians."[7] It is true: American physicians, generally speaking, have no interest in, or extensive knowledge of, nutrition.

In 1973 Ross H. Seasly of the Kettering Medical Center in Kettering, Ohio, pointed out that despite the growing body of knowledge about trace elements and their association with illness, there has been little application of research findings to clinical medicine. For example, he said, during the past ten years Kettering's clinical laboratory has not received a single request for analysis of trace metal concentrations in a patient, despite abundant evidence that such an analysis can be useful in diagnosis.[8]

The first American work on the chemical composition of foods was published by Yale scientist Wilbur O. Atwater in 1869. Atwater wrote about the chemicals in corn. At that time the measuring instruments were just becoming available to study the relationships of biochemistry to disease (it is biochemistry that provides the quantitative basis for understanding plant and animal nutrition). But historic circumstances dictated another turn of events. As a result, the branch of science that came to be called biochemistry was delayed in developing. In fact, biochemistry only came to be recognized as a distinct branch of science in 1938. Why was interest in biochemistry so slow in developing? The former president of the American Chemical Society, biochemist Roger J. Williams, said in 1971:

At the very time when the relationship between nutrition and disease might have come to the fore, medical science experienced a revolution and became intoxicated with the success of treatments that derived from Louis Pasteur's dramatic discovery [published in 1865] that microbes cause disease.

When Pasteur—a chemist, not a physician—first pro-
mulgated the novel doctrine that microbes can cause disease, he met with great resistance and the medical men of
his day would have none of his theory. It was not that
they had a better idea; it was simply that they found new
ideas distasteful. To the next generation of doctors, however, the theory of the microbial origin of disease was not
new. Spurred by what they could see through micro-
scopes and by the spectacular demonstrations of Pasteur,
Koch, and others, the medical men accepted Pasteur's
doctrine wholeheartedly—carried it farther, in fact, then
he probably would have. Pasteur had never said that *all*
diseases are microbial in origin; yet that was the assump-
tion the profession now seemed to make.[9]

Well into the twentieth century, mounting evidence showed that
nutritional deficiencies cause human diseases. Despite the evidence, medical educators have continued to heavily stress the non-
nutritional causes of disease.

Today, however, medical science thoroughly accepts the general
proposition that cell nutrition is an essential aspect of plant and animal health, even if physicians don't put this knowledge to much
practical use. As stated by E. J. Underwood in his well-known work
Trace Elements in Human and Animal Nutrition:

Continued ingestion of diets that are deficient, im-
balanced, or excessively high in a particular trace element
invariably induces changes in the functioning forms, activities or concentrations of that element in the body tissues or fluids, so that they fall below, or rise above, the
permissible limits. Under these circumstances biochemical defects develop, physiological functions are affected,
and structural disorders may arise in ways which vary
with different elements, with the degree and duration of
the dietary deficiency or toxicity, and with the age, sex,
and species of the animal involved.[10]

Despite the general acceptance of this viewpoint, medical doctors
today still receive inadequate training in nutritional health.[11] Recently a writer in the *Bulletin* of the National Health Federation
suggested why this might be so.[12] Nutritional medicine is preventive medicine. Unfortunately, the American system of compensa-

tion does not make it worth a doctor's time to practice preventive medicine; the compensation is earned by curing and alleviating disease conditions, not preventing them. In the case of doctors, compensation is a large-scale affair; the inducement *not* to study preventive medicine is formidable. The average physician in 1972 earned a net income from practice (after payment of tax-deductible professional expenses, but before payment of income taxes) of $42,700. This is income from practice alone and does not include any income from investments.[13]

As a result of the unwillingness of physicians to enter the field of nutritional medicine, this century has witnessed the creation of an information void, a seductive vacuum into which numerous food faddists have rushed. This fact in itself has tended to further discourage the serious study of nutrition; any physician who enters the field must overcome the stigma of being a suspected food faddist.[14]

Despite these drawbacks, the study of nutrition has progressed steadily throughout this century, primarily because nutritional research almost invariably pays off in increased agricultural production. Take the case of zinc. In 1869 zinc was discovered to be an essential nutrient in one particular species of fungus. Further knowledge of zinc's role in living systems had to wait thirty years more until researchers developed new measuring instruments and began discovering a range of essential organic compounds (which they called vitamins) around the time of World War I.

Recognition of the importance of essential nutrients spread rapidly during this time, and by the 1920s major discoveries were being made—some of the B vitamins, vitamin A, vitamin C, vitamin D.[15] By 1925 although no one could yet prove that zinc was an essential nutrient for higher plants, citrus growers in Florida and California were applying zinc fertilizers to their trees and realizing substantial increases in crop yield. In 1934 zinc was shown to be essential in animal nutrition. Taking considerable pains, scientists raised laboratory rats on a diet deficient in zinc; the rats never grew to full size, their hair changed color, and some of it fell out. By 1940 researchers had observed other changes in zinc-deficient rats: pathological changes in the skin, esophagus, and testicles. These results were confirmed in other laboratories. By 1955 veterinarians had found that a common disease of hogs, called parakeratosis, could be cured by adding zinc to the diet.[16] This was an important finding because it allowed hog farmers to increase production inexpensively. The well-known trace element researcher Dr. Walter J. Pories, has said:

> Zinc's importance to agriculture was recognized 30 to 40 years ago with subsequent heavy supplementation of corn, pecans, potatoes, tobacco and fruits with marked increases in yield and quality. The list of crops requiring supplementation continues to increase each year. In spite of this evident widespread zinc deficiency, few felt that the element had any relevance to animal nutrition. The discovery by Tucker and Salmon in 1954, only 15 years ago, that porcine parakeratosis was a zinc deficiency changed this attitude rapidly. Today most feeds are heavily supplemented [with zinc] with better growth, fewer abnormalities, improved feed efficiency and increased yields.
>
> Another long delay has occurred in the transference of these findings to human medicine. It is apparent that if our food crops and food animals are deficient, that man must often be deficient as well. . . . Human zinc deficiency has been recognized only during the last 7–8 years and it is now obvious that the deficit is widespread, especially in a hospital population and probably among the very young, the old, the poor, and those with various metabolic diseases.[17]

Experiments on wounded humans in 1966 showed definitely that wound healing requires zinc and in a large fraction of cases wound healing is speeded by zinc salts given therapeutically. It has since been learned that bad burns deplete the human body's zinc supplies and consequently retard healing. This may be an important reason why bad burns are usually so difficult to cure.[18] This discovery led to the recognition that many people are zinc-deficient, especially old people and sick people. It is now thought, for example, that among hospital patients "zinc deficiency is not a rare phenomenon."[19]

Zinc deficiency has now been found associated with pancreatic diabetes.[20] It has also been noted that patients suffering from atherosclerosis have very low zinc levels in their blood, and zinc therapy has brought marked improvement to significant numbers of atherosclerosis victims. Atherosclerosis is one variety of arteriosclerosis, which is a general condition known as hardening of the arteries. Among people suffering from arteriosclerosis, the walls of the blood vessels lose their resiliency, their elasticity. When this happens, the vessels may burst, or they may clog up and stop carrying enough blood through the body.

Atherosclerosis is a kind of arteriosclerosis in which the walls of the blood vessels begin to build up deposits of fatty material (such as cholesterol). As these substances build up on the walls of the arteries, the interior space of the artery naturally gets smaller and smaller until finally the artery can't carry adequate blood. When this happens, the patient is in serious trouble.

Zinc has been given as therapeutic medicine to atherosclerosis patients, and some of them have shown dramatic improvement. There seems to be evidence that atherosclerosis involves damage to the inside surface of arteries, and a zinc deficiency may prevent adequate arterial healing. We must stress that zinc is by no means the only nutrient involved in atherosclerosis; certainly chromium, vanadium, and the B vitamins are also involved, and almost certainly other nutrients as well.[21] However, the evidence points strongly to a zinc deficiency as important.

In addition to this connection with human cardiovascular disease, zinc deficiency—as noted in Chapter 3—may also be critical in the development of cadmium poisoning. In the absence of adequate zinc supplies, the body may absorb higher amounts of cadmium than usual. Once absorbed, the cadmium will compete with zinc (and with other metals) for places on enzyme-binding sites.[22] Thus zinc deficiency will increase the opportunity for cadmium to cause damage, eventually leading to chronic disease. In addition to this phenomenon, a zinc deficiency may lead to excessive absorption of Zn^{65}—a common radioactive fallout product from modern industrial weapons systems.[23]

We saw in Chapter 1 that zinc has been an essential metal for the construction of proteins since the time when life evolved in the primordial seas. We now know from recent experimental work that zinc has continued to be an essential component of well-functioning animal systems (including humans'). After it became widely recognized that zinc was an essential element in animal nutrition, the U.S. Department of Agriculture began work to discover a zinc deficiency in soils. They found such a deficiency in thirty states. More-

over, they found that the deficiency is increasing, especially in the
Western states, "almost spectacularly," according to Dr. Frank G.
Viets Jr., who has been appraising the situation for more than
twenty years.[24]

Our home state, New Mexico, is typical in what's happening to
zinc and other trace elements in the soil. Researchers at New Mexico State University at Las Cruces said recently:

> In New Mexico, iron, zinc, and perhaps manganese de-
> ficiencies are becoming prevalent. This is due in part to
> the greater demand for these nutrients because of higher
> crop yields and in part to the new fertilizer-manufactur-
> ing processes which have increased the contents of the
> major nutrients and decreased the trace contents as im-
> purities.
> . . . Although these trace elements may be needed in
> only very small quantities, they are essential and good
> yields cannot be obtained without them.[25]

Dr. Frank G. Viets Jr., who now holds the title Chief Soil Scientist
with the Agricultural Research Service, U.S. Department of
Agriculture, Fort Collins, Colorado, is the nation's leading agricultural expert on zinc in relation to soils and crops. Dr. Viets has
spent a considerable amount of time traveling throughout the western states during the past twenty years. In 1966 he wrote:

> Since 1949 . . . I have observed new instances of zinc
> deficiency affecting field and vegetable crops in states that
> had never reported zinc deficiency on field and vegetable
> crops. All but Nevada of the seventeen western states and
> Hawaii have zinc deficiency in at least one crop. . . . All
> soils are more apt to produce crops with zinc deficiency
> symptoms when pushed to high levels of production with
> other [non-zinc] fertilizers, adapted crop varieties and ir-
> rigation.[26]

In this 1966 publication Viets cites an extensive body of literature regarding the question of zinc deficiency in relation to the increasing use of phosphate fertilizers. The evidence is contradictory
because the chemical problem is exceptionally complex, but Viets
concludes that "there is no doubt that residual phosphorus from

fertilization or applied phosphorus can make zinc deficiency worse on some soils with some crops."

Viets goes on to speculate about the reasons for the spread of zinc deficiency in Western soils:

> There can be no doubt that zinc deficiency of crops is becoming more common. Since World War II, the increase has been almost spectacular. Part of this reported increase can be attributed, of course, to better trained observers aware of the symptoms characteristic of deficiency. Nevertheless, the increase in zinc deficiency is real. . . .
>
> What are the soil management factors that have changed since World War II? First, there is undoubtedly less recycling of the plant-available forms of zinc on the farm. The tractor was substituted for the horse. Fewer cattle are fed on the farm, and more of the grain and forage as hay and ensilage are sold off of the farm to a commercial feeder. When the feeder disposes of the manure, it does not return to the soil that produced the feed. . . .
>
> Second, in these two decades the use of commercial fertilizers in the United States has steadily increased from 8.5 million tons in 1940 to 20.3 million tons in 1950, to 24.4 million tons in 1960. More fertilizer should add more zinc, but the fertilizer materials have become purer—higher grade. . . .

Dr. Viets summarizes: "The most plausible explanation of the increase in zinc deficiency involves a combination of (a) less recycling of zinc on the farm; (b) actual lower incidental additions in fertilizers, and (c) much higher crop yields being obtained."

Certainly one cause of increasingly zinc-deficient soils is high-yield agricultural techniques which drive the soil harder and harder each year. We have seen the trend: output per man-hour in agriculture has increased 244 percent in just the past two decades while per-acre yield has more than doubled.[27] To achieve these increases, tremendous inputs of inorganic fertilizers (nutrients) have become standard practice. It is now well recognized that improper fertilization can cause induced deficiencies in soil. This happens when one nutrient (such as nitrogen or phosphorus) is artificially increased in soil (thus making plants grow larger and faster) and the other essential nutrients are therefore removed from the soil faster and faster too. If these other nutrients are not returned to the soil by careful, complete fertilization, imbalances must ultimately result.

Anyone who reads recent agricultural trade magazines will be struck by the change that has taken place in just the past few years. Under the regime of high-yield agriculture, the modern industrial farm is completely reliant on sophisticated electronic equipment (and expensive expertise) to maintain a nutrient balance in the soil. Along with this change in agricultural practice, there is evidence that improper fertilization is now widespread. Dr. Karl Schutte has spoken of the "four fertilizer mentality," by which he means the common belief that only four fertilizers (nitrogen, phosphorus, potassium, and calcium) are needed to maintain soil fertility.[28] This "four fertilizer mentality" has recently begun to broaden into an "eleven fertilizer mentality"; for example, a recent issue of *Agrichemical Age* lists seven elements—iron, zinc, manganese, copper, boron, molybdenum, and chlorine—as essential micronutrients.[29] Still, as we saw in Chapter 1, there are now twenty trace elements known to be essential for animal (including human) nutrition, plus four essential bulk elements (in addition to thirteen essential vitamins). Even the "eleven fertilizer mentality" is ignoring many essential elements. Table 9-1 illustrates the chemical complexity of typical soil.

Essential minerals enter into every biochemical system in plants (as they do in animals and people); for example, all of a plant's water relations (including rate of wilting, plus resistance to drought, heat, and cold) are largely governed by trace elements. In addition, the vitamin and protein content of plants is determined by the trace element content of the soils on which the plants are grown.[30] It is, therefore, of critical importance that modern farmers obtain a correct analysis of soil conditions so they can grow crops which will adequately support animal and human consumers.[31]

Other modern agricultural practices besides inorganic fertilization may induce soil deficiencies, for example, irrigation agriculture. Irrigating soil brings salts up to the surface and leaves them in the upper horizons of the soil, where they can reduce crop yields. To remove the salts, additional irrigation water must be flushed through the soil. This flushing action removes the salts, putting them into the nearest body of water, causing a salinization problem such as we find in the middle and lower Rio Grande and Colorado rivers today.[21] Perhaps ultimately more important, the flushing action also removes essential trace elements from the soil, thus requiring careful fertilization to maintain soil fertility and nutrient balance.

Other agricultural practices can also affect the trace element balance of soils. The excessive use of lime, for example, can cause at least manganese, iron and zinc deficiencies.[33] On the other hand, the use of 20 to 30 tons of manure per acre will usually cure trace element imbalances.[34] Manure contains a high percentage of organic matter which feeds microorganisms in the soil and which maintains the tilth of the soil; the microorganisms in turn help make the inorganic mineral nutrients available to the roots of plants. For this reason, the modern practice of confining cattle to feedlots (instead of letting them graze across the fields as they used to) can lead to soil deficiencies. After cattle are gathered into one place (the feedlot) for fattening, their manure is also gathered in the one place. From here, instead of returning to the soil, feedlot manure is processed—when it is processed at all—much like municipal (human) sewage: first treated to remove odor and kill bacteria (though not viruses), then dumped in the nearest body of water. Or it is placed in land-fill dumps. In place of the nutrients that used to be naturally returned to soil in manure, modern farmers are, as we have seen, dependent on sophisticated knowledge (plus capital equipment and operating expertise) to maintain soil fertility using inorganic fertilizers. On top of that, some of the land that used to be grazed by cattle is now turned to intensive grain farming, to provide feed for the feedlots; thus the old grazing lands are now intensively driven, thus requiring even further applications of inorganic fertilizers. Altogether, as Paul and Anne Ehrlich have said, modern agriculture worldwide is an "ecological disaster area." This modern combination of practices is leading more and more to very serious soil depletion and nutrient imbalances. It has been proposed that zinc supplementation of grain diets now might improve the growth and well-being of large segments of the human population.[35]

It has now been confirmed that a zinc deficiency has been found in a large group of Denver, Colorado, children of well-to-do white parents.[36] Researchers at the Department of Pediatrics, University of Colorado Medical Center, have reported measuring zinc in the hair

Table 9-1. The chemical composition of typical soil.

Chemical symbol	Element	Average parts per million in dry soil. The range is given inside parentheses.	
Ag *	silver *	0.1 *	(0.01 to 5)
Al	aluminum	71,000	(10,000 to 300,000)
As	arsenic	6	(0.1 to 40)
B	boron	10	(2 to 100)
Ba	barium	500	(100 to 3,000)
Be *	beryllium *	6 *	(0.1 to 40)
Br	bromine	5	(1 to 10)
C	carbon	20,000	
Ca	calcium	13,700	(7,000 to 500,000)
Cd *	cadmium *	0.06 *	(0.01 to 0.7)
Ce *	cermium *	50 *	
Cl	chlorine	100	
Co	cobalt	8	(1 to 40)
Cr	chromium	100	(5 to 3,000)
Cs *	cesium *	6 *	(0.3 to 25)
Cu	copper	20	(2 to 100)
F	fluorine	200	(30 to 300)
Fe	iron	38,000	(7,000 to 550,000)
Ga	gallium	30	(0.4 to 300)
Ge *	germanium *	1 *	(1 to 50)
Hf	hafnium	6	
Hg *	mercury *	0.03 *	(0.01 to 0.3)
I	iodine	5	
K	potassium	14,000	(400 to 30,000)
La *	lanthanum *	30 *	(1 to 5,000)
Li	lithium	30	(7 to 200)
Mg	magnesium	5,000	(600 to 6,000)
Mn	manganese	850	(100 to 4,000)
Mo	molybdenum	2	(0.2 to 5)
N	nitrogen	1,000	(200 to 2,500)
Na	sodium	6,300	(750 to 7,500)
Ni	nickel	40	(10 to 1,000)
O	oxygen	490,000	
P	phosphorus	650	
Pb	lead	10	(2 to 200)
Ra	radium	8×10^{-7}	(3 to 20×10^{-7})
Rb	rubidium	100	(20 to 600)
S	sulfur	700	(30 to 900)
Sb *	antimony *		(2 to 10 ?) *
Sc	scandium	7	(10 to 25)

(Continued on following page)

of a large group of children and correlating this with objective measurements of (a) the ability to taste, (b) size and weight of child, and (c) loss of appetite. Children with low amounts of zinc in their hair (a reliable indicator of the body's total zinc burden) were considerably smaller and less well developed than children with average amounts of zinc in their hair. Zinc supplementation of the diet brought measurable increases in total food intake (return of the lost appetite) and increased ability to make taste discriminations. The researchers concluded:

> There is evidence that a substantial number of adults in the United States do not have optimal zinc nourishment, and that the content of zinc in the average diet is not sufficient to guarantee an adequate intake at times of increased requirement. . . .
> Results of the present studies indicate that zinc deficiency may occur in otherwise normal children in this country. Moreover, levels of zinc in hair indicate that low stores of zinc in the body are common in infants and young children.[37]

Chemical symbol	Element	Average parts per million in dry soil. The range is given inside parentheses.	
Se	selenium	0.2	(0.01 to 2)
Se	silicon	330,000	(250,000 to 350,000)
Sn *	tin *	10 *	(2 to 200)
Sr	strontium	300	(50 to 1,000)
Th	thorium	5	(0.1 to 12)
Ti	titanium	5,000	(1,000 to 10,000)
Tl *	thallium *	0.1 *	
U *	uranium *	1 *	(0.9 to 9)
V	vanadium	100	(20 to 500)
Y	yttrium	50	(25 to 250)
Zn	zinc	50	(10 to 300)
Zr	zirconium	300	(60 to 2,000)

* Indicates that values have been established tentatively and may be subject to revision in future.

Table 9-1 adapted from H. J. M. Bowen, **Trace Elements in Biochemistry** (New York: Academic Press, 1966), pp. 39-40.

Another trace element which almost certainly has widespread important effects on the health of Americans is the familiar shiny metal chromium. It has only been since 1954 that scientists have suspected any biological activity for chromium.[38] However, during the past nine years a great deal of research activity has shown that chromium is essential for human health[39] and that millions of Americans probably don't maintain an adequate supply in their bodies. Although Dr. Henry Schroeder's laboratory was by no means the only source of new information on chromium, still it was Dr. Schroeder's work which showed most dramatically that a chromium deficiency causes a disease very much like diabetes mellitus in laboratory animals.[40] Here was unmistakable evidence that this rather obscure trace element had direct, critical effect on animal health. An estimated 4,400,000 Americans were suffering from diabetes in 1968 (the last year for which figures are available to us). Of these, an estimated 1,600,000 didn't even know they had it. Diabetes was the direct cause of death in 35,049 cases in 1967; in addition, diabetes was listed as a major contributory cause of death in an estimated 51,000 more cases.

Most important, diabetes mellitus has been killing a steadily larger percentage of Americans since about 1960, when an upward trend was first noted. Most striking is the difference in rates of increase between different racial groups. Deaths from diabetes among white people increased from 16.4 deaths per 100,000 population in 1950 to 17.3 per 100,000 in 1967. During the same time period, however, deaths from diabetes among nonwhites skyrocketed from 14.4 in 1950 to 21.1 in 1967. These figures indicate an increase in the diabetes death rate of 5 percent among whites and 47 percent among nonwhites during the eighteen-year period.[41]

When Dr. Schroeder and his colleagues discovered that chromium deficiency could cause a disease like diabetes in rats and mice, they started measuring chromium in human beings to see whether they could detect this metal or not. They did manage to detect it, and to their surprise, they found that the body burden of chromium continually declines in Americans after birth. No other element consistently follows this pattern. American babies are born with adequate amounts of chromium, mothers' bodies being very efficient at transferring their chromium supply into the growing fetus, which needs it. However, after birth the body burden of chromium declines "precipitously" for the first two or three decades of life and then continues to decline slowly but steadily thereafter.[42]

Dr. Schroeder and scientists in several other laboratories have tried to figure out why Americans lose their chromium. They have noted that people from other parts of the world—Africa, for example, and the Middle East—don't generally lose their body burden of chromium the way Americans do. A survey of human tissues from various countries showed no detectable chromium in organs of 23.3 percent of middle-aged Americans; only 1.6 percent of same-aged foreigners had equally depleted chromium stores.[43]

Modern industrial food processing is partly responsible for chromium losses (and for other deficiencies). For example, the nearly universal practice of modern wheat milling (to remove the germ and bran, which tend to spoil on the shelf) results in removal of the following elements: 60 percent of the calcium, 75 percent of the phosphorus, 77 percent of the potassium, and 77 percent of the sodium originally present. In addition, modern wheat milling removes these amounts of essential trace nutrients: 86 percent of the manganese, 76 percent of the iron, 89 percent of the cobalt, 85 percent of the magnesium, 68 percent of the copper, 78 percent of the zinc, 48 percent of the molybdenum, and 50 percent of the chromium. Modern food processors remove all these essential mineral nutrients, then add iron to their white flour and fraudulently sell it as "enriched."[44]

There is a possibility that heating food may convert chromium in vegetables to a chemical form which the body can't use readily.[45] However, even these losses of dietary chromium appear to be inadequate to explain why Americans lose their chromium stores. Dr. Schroeder has, however, proposed an answer for this puzzle. He has pointed to several laboratory experiments which show that eating refined sugar causes people to excrete large amounts of chromium in their urine.[46]

The mechanism seems to be this: Introducing sugar into the bloodstream rapidly calls forth insulin and chromium into the blood, which work together in metabolizing the energy-rich sugar. During this process the kidney captures and excretes about 20 percent of the chromium that is in the blood.

Dr. Schroeder has calculated that Americans eat an average of 207 grams of grains[47] each day and at least 140 grams (5 ounces) of refined sugar.[48] Calculated on the basis of the sugar alone, an average American diet provides ample opportunity for the body's total burden of chromium to be excreted in a decade or two.

This is an unnatural condition created mostly, it appears, by modern food-processing methods. In their natural state, foods ordinarily contain the trace elements which are necessary for their digestion by people. However, as we have indicated, modern grain-milling methods and the refinement of sugar happens to remove the bulk of the trace elements from those foods.

As a result, these highly refined foods do not contain the nutrients necessary for the body to use when metabolizing the foods to extract their energy. The result after decades of eating these foods is severely depleted chromium stores and a growing inability to metabolize glucose. This is a form of diabetes.

The human body must have chromium, or it loses its ability to handle both sugars and fats (lipids) which can then pile up in the bloodstream and can cause diabetes and atherosclerosis.[49] Diabetes is a disease of sugar metabolism; atherosclerosis is a disease of fat metabolism. These diseases are closely related; in fact, a high percentage of diabetics eventually develop atherosclerosis. Atherosclerosis, we have noted earlier, is one aspect of arteriosclerosis (hardening of the arteries). In atherosclerosis, the walls of the blood vessels keep receiving deposits of materials like cholesterol, a fatty substance normally found in the bloodstream in small amounts. Excessive amounts of cholesterol cause a buildup on the inside walls of the blood vessels, and as a result, the blood vessels end up with a reduced carrying space for blood. One result can be an inadequate blood supply to many parts of the body. When atherosclerosis thickens the arteries feeding blood to the heart, then a blood clot may get stuck in the reduced passageway; if that happens, the heart begins to starve for lack of oxygen-rich blood. At this point the person suffers a heart attack, resulting in permanent injury to the heart or death.

When these facts about cholesterol and atherosclerosis made headlines in the late 1950s and early 1960s, millions of Americans became very diet-conscious and tried to eat foods low in cholesterol. What these people didn't realize is that most of the cholesterol absorbed from the diet, like every other dietary compound, is broken down during metabolism and will be re-formed into cholesterol when the body has a need. Your body can manufacture cholesterol from other substances whether you eat much cholesterol or not.

Cholesterol is a normal constituent of the body. The brain and pancreas are up to 5 percent (50,000 ppm) cholesterol and the liver is up to 1 percent pure cholesterol. Without cholesterol the skin would dry up, the brain would not function, and you would have no adrenal or sex hormones.[50] The body regulates the manufacture of cholesterol by a mechanism which depends on the presence of at least three trace metals—chromium, manganese, and vanadium—and probably on others as well.[51] If these trace metals are lacking or are present in unbalanced amounts, serious disturbances of glucose and lipid metabolism will occur. This has now been clearly established.[52] Many other abnormal conditions besides chromium deficiency may cause impaired glucose or lipid metabolism.[53] No one claims that chromium deficiency causes all diabetes or all atherosclerosis. However, chromium salts given as therapeutic medicines under controlled conditions have resulted in significant improvement in significant numbers of human patients. Some people suffering from diabetes and some suffering from atherosclerosis have been definitely improved by chromium therapy, it has been reported, within the last five years. As R. J. Doisy and others have pointed out, we can't expect chromium therapy to improve any disease condition except one which a chromium deficiency specifically causes. If a patient's diabetic or atherosclerotic condition is caused by several factors, chromium deficiency being only one of them, then only a slight improvement, or no improvement at all, can be expected from chromium therapy.

In experiments with three different groups of diabetics, chromium therapy brought measurable improvement to 40 percent, 67 percent and 33 percent.[54] This is an impressive demonstration of the potential importance of chromium and other trace metals in human disease. Dr. Schroeder is now so convinced of the importance of chromium (and vanadium) that he flatly predicts that one day soon atherosclerosis will be prevented, if not cured, by trace metal therapy.[55]

It has long been thought that diabetes is strictly an inherited disease, a genetic defect which shows up as a lack of one particular enzyme. However, Dr. Schroeder has produced clinical symptoms resembling diabetes in five successive generations of laboratory animals by inducing a chromium deficiency.[56] Schroeder points out that if he had not intentionally induced a chromium deficiency in the five generations, he—like any other researcher—would have thought he was observing a hereditary disease. Among people, there is evidence that diabetes strikes the poor much more frequently than the wealthy.[57] Since our best source of dietary chromium is animal protein, which is expensive food, one wonders whether all diabetes is simply inherited or whether it's partly a widespread deficiency disorder.

In addition to relieving the symptoms of diabetes and preventing the development of atherosclerosis in experimental animals (and in people), chromium fed to rats and mice causes them to live much longer than a control group.[58] This is an important finding which confirms that many of the body's systems will function *partially* (at reduced efficiency) under conditions of malnutrition. Millions of man-years (and woman-years) are lost in this nation alone each year because people lose their capacities or die needlessly early. At a time when the world has never had greater need of a fund of accumulated knowledge and wisdom, the premature loss of our middle-aged to elderly group represents an extravagant waste and a burdensome cost to us all. Biologically, they are the educational and therefore adaptive resource that has made cumulative human civilization possible; their increasing loss will be felt most keenly generations hence.

> Evidence that diets in the United States have become worse since 1955 was reported in the preliminary report of the USDA [U.S. Department of Agriculture] 1965 survey and is supported by other studies. . . . [O]ur review of the individual dietary intake studies published between 1960 and 1968 indicates that in recent years diets may have changed for the worse for all [seven] nutrients studied.[59]
>
> —Thomas R. Davis and others (1969)

10

Do You Need a Weatherman to Know Which Way the Wind Blows?

Deteriorating Environmental and Human Health

GENERAL environmental deterioration has now been firmly established as a cause of human disease and death. For example, in urban areas of Japan, Sweden, Britain, and the United States, numerous studies have now confirmed that air pollution is causing increases in bronchitis, emphysema, lung cancer, stomach cancer, and Asian influenza among others.[1] On top of that, there is sound evidence relating air pollution levels to increases in the total human death rate, the death rate among fetuses and newborn infants, and the death rate among young children. In addition, recent studies in New York City have shown that as pollution levels rise, old people start dying with unusual frequency from heart disease and respiratory disorders. (The New York studies have emphasized that death rates rise even under moderate pollution conditions; it is not necessary to wait for the rare or unusual pollution event before observing excess deaths.[2]) The International Agency for Research in Cancer (IARC) has recently announced its estimate that 75 to 85 percent of all human cancers are caused by environmental factors, chiefly radiation and chemicals.[3] The President's Committee on Mental Retardation, noting in January 1972, that, conservatively estimated, some 6 million Americans now suffer from mental retardation, said that "elements of the environment not only affect their lives but often are the agents that cause them to be retarded."[4]

169

An estimated 25 percent of America's hospital beds and beds in other institutions are filled by people suffering from mental and physical illnesses thought to be caused by genetic disorders. An estimated one out of every fifteen children is now born with a major birth defect.[5] The costs of birth defects are very high.

Disease statistics and death rates in recent years suggest that the environmental health of average Americans has been deteriorating for a decade; in fact, statistics from a growing list of industrialized nations indicate that people are living shorter lives now than they did a decade ago.[6] Take the case of a typical sixty-year-old man. The typical American male born in the year 1780 when he reached age sixty could expect to live an average of 14.8 more years. One hundred and fifty years later, in 1930, when the nation was fully industrialized, a sixty-year-old man could expect to live only 14.3 more years. It seems that industrialization had begun in the early 1930s to shorten the lives of average American men.[7] Then the discoveries of sulfa drugs in the mid-1930s and antibiotics in the late 1930s and early 1940s increased the life expectancy of the sixty-year-old man by 1.7 years.[8] Now however the trend is once again downward. In the period 1959–1961, a sixty-year-old white male could expect to live 16.0 more years (whereas a nonwhite male could expect to live only 15.3 more years). However, in 1968 the years of life remaining for a sixty-year-old white male had dropped to 15.8 years, and for nonwhite males the figure had dropped to 14.5 years. (See Table 10-1.)

This downward trend in longevity, noticeable among U.S. white and nonwhite males, has not occurred among the female populations of America, white or nonwhite. Women continue to live a little longer each year. In only one country that we know of, Australia, is there evidence that women are living shorter lives than they used to.[9]

A general deterioration of health can also be observed in examining Table 10-2 which shows the death rate for U.S. populations, by sex and by skin color, 1961 through 1968. The worsening of the death rate is most clearly evident among nonwhite males, then among white males, then among nonwhite females (where the numbers are equivocal). Among white females the death rate has consistently been dropping since 1961, except in 1967–1968, when it rose slightly.

Table 10-1. Average remaining lifetime, in years, at specified ages, for U.S. white and nonwhite males, 1959-1961 and 1968.

Age	Nonwhite males		White males	
	1959-1961	1968	1959-1961	1968
0 . . .	61.5	60.1	67.6	67.5
1 . . .	63.5	61.4	68.3	68.0
5 . . .	60.0	57.8	64.6	64.3
10 . . .	55.2	53.0	59.8	59.4
15 . . .	50.4	48.2	54.9	54.6
20 . . .	45.8	43.6	50.3	49.9
25 . . .	41.4	39.4	45.7	45.4
30 . . .	37.1	35.3	41.0	40.8
35 . . .	32.8	31.2	36.3	36.1
40 . . .	28.7	27.4	31.7	31.6
45 . . .	24.9	23.7	27.3	27.2
50 . . .	21.3	20.3	23.2	23.0
55 . . .	18.1	17.2	19.5	19.2
60 . . .	15.3	14.5	16.0	15.8
65 . . .	12.8	12.1	13.0	12.8
70 . . .	10.8	10.5	10.3	10.2

Table 10-1 adapted from U.S. Department of Health, Education and Welfare, **Facts of Life and Death** [Public Health Service Publication No. 600, Revised 1970] (Washington, D.C.: U.S. Government Printing Office, 1970), p. 28.

Through this century, dramatic reductions in the disease and death rates from bacterial infections, made possible by the discovery of sulfa drugs and antibiotics, have been countered on the other side of the picture by steady increases related to environmental insult (especially lung diseases) and to diet (atherosclerosis, hypertension, and related heart diseases, cancer, diabetes). Table 10-3 shows the fastest-growing diseases in America between 1900 and 1967; they are heart disease, all cancers (most of the increase being lung cancers, though cancers of the colon and rectum are increasing dramatically too), miscellaneous diseases of the blood circulation system, and miscellaneous diseases of the respiratory system (lungs and throat). Table 10-4 shows that cancer, diabetes mellitus, and cirrhosis of the liver are increasing quite rapidly today. Table

Table 10-2. U.S. age-adjusted death rate per 1,000 population, 1961 through 1968.

| Year | Nonwhite | | White | |
	Female	Male	Female	Male
1961	8.6	11.6	5.4	8.9
1962	8.7	12.0	5.4	9.0
1963	8.9	12.5	5.5	9.2
1964	8.6	12.2	5.3	9.0
1965	8.5	12.4	5.3	9.1
1966	8.6	12.7	5.3	9.2
1967	8.2	12.4	5.2	9.0
1968	8.6	13.3	5.3	9.2

Table 10-2 adapted from U.S. Department of Health, Education and Welfare, **Facts of Life and Death** [Public Health Service Publication No. 600, Revised 1970] (Washington, D.C.: U.S. Government Printing Office, 1970), p. 10.

Table 10-3. Death rates per 100,000 population for important causes of death which increased between 1900 and 1967, including percent of increase.

Cause of death	1900	1967	Percent increase, 1900 to 1967
Diseases of heart	137.4	364.5	165.3
Malignant neoplasms (cancer)	64.0	157.2	145.6
Miscellaneous diseases of circulatory system	12.0	15.1	25.8
Miscellaneous bronchopulmonary diseases	12.5	14.8	18.4

Table 10-3 adapted from U.S. Department of Health, Education and Welfare, **Facts of Life and Death** [Public Health Service Publication No. 600, Revised 1970] (Washington, D.C.: U.S. Government Printing Office, 1970), p. 12.

Table 10-4. 1970 U.S. death rates as a percentage of 1969 rates and as a percentage of 1965-1969 average, for selected causes.

Cause of death	Death rate per 100,000 population in 1970	1970 death rate as a percentage of 1969 death rate	1970 death rate as a percentage of 1965-1969 death rate
Cancer (all forms)	161.9	101	103
Diabetes mellitus	18.6	101	104
Cirrhosis of the liver	15.5	103	114

Table 10-4 adapted from "Current Mortality Report," **Statistical Bulletin** [of Metropolitan Life Insurance Co.], Vol. LII (February, 1971), p. 11.

Table 10-5. U.S. cancer death rates, by sex and by skin color, 1940 through 1960, per 100,000 population.

	Nonwhite					White				
	1940	1945	1950	1955	1960	1940	1945	1950	1955	1960
Women	129	127	141	140	136	143	139	132	128	121
Men	94	104	138	160	174	138	142	148	157	159

Table 10-5 adapted from National Cancer Institute, **1971 Fact Book** (Bethesda, Md.: National Cancer Institute, National Institute of Health, February, 1971), p. 12.

10-5 shows which populations in the United States are hardest hit by all cancers. Up until 1955 nonwhite men used to have less cancer than white men. After 1955 the picture changed, and the incidence of cancer among nonwhite men surpassed the incidence among white men; since 1955 the situation for nonwhite men has continually worsened. The forced urbanization of black people has doubtless taken its toll.

In May 1972, Howard University doctors announced finding an "alarming" increase in cancer among black people. Drs. Ulrich K. Henschke and Jack E. White, representing the Howard research team, said that greater exposure to cancer-causing substances in the environment of blacks "must be suspected as the main cause."[10] Among nonwhite women the cancer rate increased dramatically between 1945 and 1950 but has declined steadily ever since, though it remained higher through 1968 than it had been in 1940. Among white women the cancer rate has been steadily dropping since 1940. Among white men, on the other hand, the rate of cancer has steadily increased since 1940. In 1940 women of all skin colors had more cancer than men; by 1945 among white people and by 1955 among nonwhite people, the situation was reversed, and women had less cancer than men. Cancer is the Number Two killer among all Americans.

The Number One killer, of course, is heart disease. Table 10-3 shows that heart disease has been the fastest-growing disease during this century (though lung diseases have been faster-growing in the past decade; see Tables 10-6 and 10-7). Table 10-8 shows that heart disease is the chronic disorder most often responsible for physical disability among Americans. As we can see in Table 10-9, heart disease, stroke, hardening of the arteries, and other diseases of the circulatory system together kill 56 to 57 percent of all Ameri-

---

Table 10-6. U.S. death rates for malignant neoplasms (cancer) of the lung, 1950 and 1967, including percentage of increase.

	1950		1967		Percent increase, 1950 - 1967	
	male	female	male	female	male	female
Nonwhite	6.5	1.6	22.6	4.3	248	169
White	11.9	3.1	28.8	6.0	142	94

Table 10-6 adapted from U.S. Department of Health, Education and Welfare, **Facts of Life and Death** [Public Health Service Publication No. 600, Revised 1970] (Washington, D.C.: U.S. Governing Printing Office, 1970), p. 23.

Table 10-7.   Death rates per 100,000 U.S. population for specified chronic diseases of the respiratory system, 1950 and 1967, including percentage of increase.

Cause of Death	1950	1967	Percent increase, 1950 to 1967
Bronchitis	2.0	3.2	60.0
Chronic bronchitis	.8	2.3	187.5
Bronchitis with emphysema	.2	1.8	800.0
Other diseases of lung and pleural cavity	1.2	12.0	900.0
Emphysema without mention of bronchitis	.8	10.6	1225.0

Table 10-7 adapted from U.S. Department of Health, Education and Welfare, **Facts of Life and Death** [Public Health Service Publication No. 600, Revised 1970] (Washington, D.C.:   U.S. Government Printing Office, 1970), p. 25.

cans who die each year. Cancers kill another 17 to 18 percent. Between them, cardiovascular diseases and cancers kill three out of every four Americans who die each year (73.2 percent).

In addition to these signs of deteriorating health, there are others as well. For example, in the period 1958 to 1967, the age-adjusted death rate for emphysema rose from 5.8 to 14.9 per 100,000, an increase of 157 percent in ten years. During the same period, there was a reported 90 percent increase in the death rate caused by bronchitis, the death rate rising from 1.9 to 3.6 during the period.[11] Table 10-10 shows the rising incidence of leukemia (cancer of the blood cells) in the United States, 1950 to 1967 (a 22 percent increase during the period).

In Table 10-11 we can see that, for hardening of the arteries (arteriosclerosis) and for degenerative heart disease, the United States has an exceptionally high death rate when judged against other industrialized nations. Only Finland has a higher death rate from these diseases.

Table 10-8. Ten leading chronic health conditions causing limitations on activities among U.S. population during the period 1965-1967.

Condition causing limitation of activities	Estimated number of cases
Heart conditions	3,600,000
Arthritis and rheumatism	3,248,000
Impairments (except paralysis) of back or spine	1.796,000
Mental and nervous conditions	1,711,000
Impairments (except paralysis or absence) of lower extremities and hips	1,351,000
Visual impairments	1,222,000
Hypertension without heart involvement . . .	1,187,000
Asthma-hayfever	1,065,000
Paralysis (complete or partial)	925,000
Conditions of genitourinary system	894,000

Table 10-8 adapted from U.S. Department of Health, Education and Welfare, **Facts of Life and Death** [Public Health Service Publication No. 600, Revised 1970] (Washington, D.C.: U.S. Governing Printing Office, 1970), p. 33.

This table shows conditions limiting major activities such as work (both inside and outside the home) and going to school as well as conditions causing lesser limitations not related to major activity. Not included in these figures are conditions causing limitations among persons in resident institutions (sanitariums, chronic disease hospitals, nursing homes, and homes for the aged).

Finally, though this is probably a general measure of the "quality of life" rather than of specific disease conditions, the suicide rate in the United States has increased steadily since 1955. In 1955 the suicide rate was 1.5 per 100,000 population; by 1967 it had nearly doubled (to 2.7 per 100,000), bringing the suicide rate back up to what it had been during 1935 in the depths of the Great Depression.[12]

All these numbers about human death and disease make one strong point: We are slowly but relentlessly destroying our own habitat. As a writer for *Scientific American* recently put it, "new signs of stress on the biosphere are reported almost daily. . . . [and human activities] are producing ominous alterations in the biosphere, not just on a local scale, but for the first time in history, on a global

Table 10-9. Ten leading causes of death among U.S. population, 1967, by sex.

Rank	Cause of death	Percent of all deaths from this cause	Death rate per 100,000 people from this cause
	FEMALE		
1	Diseases of the heart	37.9	301.9
2	Malignant neoplasms (cancer)	17.6	140.2
3	Vascular lesions affecting central nervous system (stroke)	13.6	107.9
4	Accidents	4.4	34.9
5	Influenza and pneumonia (except pneumonia of the newborn)	3.1	24.7
6	General arteriosclerosis (hardening of the arteries)	2.6	20.7
7	Diabetes mellitus	2.6	20.4
8	Certain diseases of early infancy	2.5	19.8
9	Other diseases of the circulatory system	1.5	12.2
10	Cirrhosis of liver	1.2	9.8
	Other causes	13.0	
	MALE		
1	Diseases of the heart	39.8	430.1
2	Malignant neoplasms (cancer)	16.2	174.9
3	Vascular lesions affecting central nervous system (stroke)	13.6	107.9
4	Accidents	4.4	34.9
5	Influenza and pneumonia (except pneumonia of the newborn)	3.0	33.0
6	Certain diseases of early infancy	2.7	29.2
7	Miscellaneous bronchopulmonary diseases	2.3	24.4
8	Cirrhosis of the liver	1.7	18.5
9	Other diseases of the circulatory system	1.7	18.2
10	General arteriosclerosis (hardening of the arteries)	1.6	17.2
	Other causes	13.0	

Table 10-9 adapted from U.S. Department of Health, Education and Welfare, **Facts of Life and Death** [Public Health Service Publication No. 600, Revised 1970] (Washington, D.C.: U.S. Government Printing Office, 1970), p. 13.

178

Table 10-10. Number of leukemia cases per 100,000 population in the U.S., 1950 to 1967. (Leukemia is cancer of the blood-forming cells.)

1950	1955	1960	1965	1967
5.9	6.5	7.1	7.0	7.2

Table 10-10 adapted from U.S. Department of Health, Education and Welfare, **Facts of Life and Death** [Public Health Service Publication No. 600, Revised 1970] (Washington, D.C.: U.S. Government Printing Office, 1970), p. 22

scale as well."[13] Air pollution, water pollution, and soil pollution now threaten the capacity of the planet to sustain future human activity (and perhaps all biological activity). As the director of graduate studies in biology at Stanford University wrote in 1970, ". . . it is possible that the capacity of the planet to support human life has [already] been permanently impaired."[14]

Air pollution facts alone indicate serious large-scale changes in the biosphere. Careful measurements of the air in the northern hemisphere in 1907 and again in 1967 have shown that atmospheric turbidity (dustiness or haziness) has increased by a factor of 2 during this century.[15] More recently, in the rural regions of America the number of particles found in an average sample of air increased by 12 percent between 1962 and 1966.[16] Summarizing the situation in 1971, the President's Council on Environmental Quality said:

> With low levels of economic activity and a sparse population, wastes could be assimilated with only minor damage to the environment. That is no longer the case. The capacity of rivers, lakes, and the atmosphere to absorb current waste loads is severely taxed. And pollution has, in many cases, reached intolerable levels.
> The physical, chemical, and biological quality of the air, water, and the land have been altered on a massive scale.[17]

There is a lot of serious speculation and discussion going on among meteorologists about the long-run global effects of human

Table 10-11. Death rates from arteriosclerosis (hardening of the arteries) and degenerative heart disease, males aged forty-five through fifty-four, for selected countries, 1967, per 100,000 population.

Finland	468.6	West Germany	179.7
United States	351.8	Czechoslovakia	178.1
Australia	325.0	Hungary	160.7
Canada	314.7	Venezuela	139.6
United Kingdom	244.0	Italy	127.5
Israel	202.9	Switzerland	126.2
Norway	200.2	Yugoslavia	80.7
Netherlands	191.2		

Table 10-11 adapted from "Cardiovascular and Chronic Respiratory Diseases: Some Data on the Magnitude of the Problem," in typescript (Bethesda, Md.: National Heart Lung Institute, October, 1970), p. 3.

---

activities. For example, as a direct result of industrial activity (burning fossil fuels and cutting down forests) the amount of carbon dioxide in the world's atmosphere is increasing.[18] Between 1958 and 1968, CO_2 increased at a steady 0.7 ppm each year; in 1968 and 1969 the annual rate of increase nearly doubled, to 1.35 ppm per year.[19] No one really knows why the rate of change changed. Total human additions to atmospheric CO_2 have already been remarkable: Between 1860 and the present, the amount of CO_2 in the earth's atmosphere increased from 290 ppm to 325 ppm—an increase of just over 12 percent. In addition to global increases in carbon dioxide and regional increases in turbidity, there has apparently been a recent increase in the number of cirrus clouds (filmy, fleecy clouds at about 33,000-foot altitude) in the sky.[20]

What are the effects of increasing the carbon dioxide content of the atmosphere, intensifying the general haziness of the atmosphere, and increasing the number of cirrus clouds? No one can say with certainty because the world's weather systems aren't well understood. However, it is known that the temperature of the earth—which averages a constant 58° Fahrenheit—depends on how much sunlight the planet absorbs and how much sunlight is reflected back into space. It is also known that turbidity, CO_2, and cloudiness can change the albedo (reflectivity) of the planet, making it a certainty that from CO_2 alone human activities will sooner or later change the temperature of the earth and affect the climate on a regional and global scale. It has been proposed by well-known scientists that if we continue our present patterns, we will most likely increase the atmosphere's CO_2 content into the range of 375 or 400 ppm (an increase ranging from 29 percent up to 38 percent) by the year 2000.[21] This might raise the average temperature of the planet as much as 1° Centigrade (1.8°F). This could lead to the melting of the polar ice caps. If this happened, the oceans would rise some 150 feet or more, inundating most of the world's cities and submerging vast tracts of prime agricultural land. (Facing is a map showing how the shoreline of the United States would change if the polar ice caps were to melt.)

On the other hand, the combustion of fossil fuels which adds CO_2 to the air is also introducing large quantities of other gases into the atmosphere and these ultimately turn into tiny particles. Combustion also directly creates substantial quantities of very fine particles. These tiny particles themselves are a very serious health hazard to people because our lungs cannot protect themselves against particles smaller than 1 micron in diameter.[22] Furthermore, these small particles are not controlled by existing air pollution abatement technology.

In addition to penetrating lungs efficiently, submicron particles are especially efficient at reflecting light. They cause haze, or turbidity. Increasing turbidity will increase the albedo of the planet. This could lower the average temperature of the earth. According to some scientists, merely dropping the temperature of the planet 3° to 7° Fahrenheit could increase the size of the polar ice caps. As the ice caps grew, creeping toward the temperate latitudes, the ice caps themselves might change the albedo of the earth, making the world still cooler. A permanent global ice cover could result. [23]

As we write, several very large climatic developments are being reported, including drought in sub-Sahara Africa and elsewhere,[24] drought in parts of the United States,[25] and an increase in the size of the global ice cap by 12 percent in 1972,[26] reported by satellite observation. The ice cap is now reportedly holding steady at its new size. Not many scientists are willing to speculate publicly on the causes of these phenomena, but one well-known meteorologist, Dr. Reid A. Bryson, director of the Institute for Environmental Studies at the University of Wisconsin, concludes that the earth's climate is changing for the worse (from the viewpoint of global food produc-

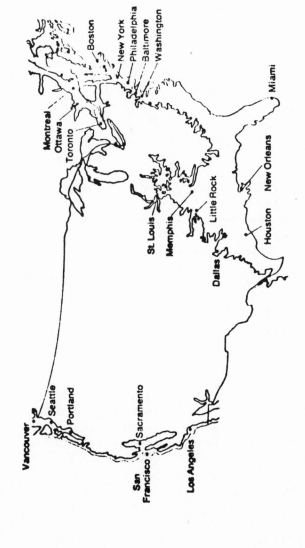

Figure   10-1

Figure 10-1 reproduced from Richard Wagner, *Environment and Man* (New York: W.W. Norton, 1971), p. 152.

tion) and that industrialization is probably a major contributing cause to the changes apparent everywhere.[27]

Citing reports by the National Oceanographic and Atmospheric Administration (NOAA), Bryson hypothesizes that carbon dioxide and small particulates in the air, added to the already-large volumes of gases and particles from natural volcanoes, are increasing the earth's albedo. The planet would therefore be reflecting more sunlight than it used to, and absorbing less solar energy.

Bryson believes that by cooling the planet slightly (by increasing its albedo), human activities have already made the swirling masses of polar air larger. As a result, says Bryson, the world's deserts are being pushed toward the equator, and the countries dependent for water monsoons are experiencing droughts. There is a possibility that the monsoons will never return. This would be a major disaster.

There is no doubt that major deserts *are* expanding toward the equator, probably partly owing to climatic changes but partly owing also to deforestation and to overgrazing by domestic livestock owned by hungry people. During this century the percentage of the earth's surface which is desert has doubled, from 10 to 20 percent.[28] People have definitely become a force capable of affecting regional and even global climate. The implications are starkly ominous.

The Ford Foundation's Energy Policy Project concluded in 1974 that "a point of no return" with regard to climatic change may be reached in as few as thirty years or as many as one hundred or longer.[29] As we've seen, there's evidence that even this estimate may be overly optimistic.[30]

Besides changing the weather, human industrial activity is having other impressively destructive impacts on a global scale. At least half a million different pollutants now enter the oceans each year. Oil alone now reaches the oceans at the rate of 6 million metric tons per year. Reporting this fact in 1974, Dr. E. Bright Wilson felt constrained to add, "The ocean is not dead—or not dead yet."[31]

The shorelines of most of the world are severely and increasingly polluted, especially the estuaries. Estuaries are the places where rivers meet the sea. Here, where the water varies in salt content and where nutrient supplies are rich, up to 90 percent of all commercially valuable fish species spend part of their lifecycle.[32] We tend to think of the world's oceans as being so immense that we could not possibly harm them. But this is not correct because of the physical characteristics of ocean ecosystems. Although the oceans are very large—139,545,000 square miles of total surface area, with a volume of 359 quadrillion (3.59×10^{17}) gallons—we tend to put our pollution directly into the most productive parts of the oceans, the coastal regions. These areas receive oil spills, fallout from air pollution, and river outflows bearing pesticides, fertilizers, heavy metals, and a myriad of other exotic industrial chemicals.

Ninety-nine percent of the ocean's productivity occurs beneath only 10 percent of the ocean's area. Fully half of the ocean's annual productivity occurs in coastal upwellings amounting to only 0.1 percent of the ocean's surface area.

The ocean is vast, but most of it is biologically a vast desert. Its productive parts are precisely the places where human pollution is occurring most heavily.[33] As a consequence, there is considerable evidence that we are well along toward poisoning our entire oceanic food supply. Put another way, we are poisoning the resource that now supplies the human species with 10 to 20 percent of its total animal protein intake.[34] The folly of this in a protein-starved world must be obvious to everyone.

A 1972 report in *Environmental Science and Technology* indicates that fish protein concentrate—one of the miracle foods that was supposed to help alleviate the world's impending food shortage—contains a concentration of toxic mercury five times higher than the fish from which the protein concentrate derived.[35] A factor of 5 increase over natural background levels (0.2 ppm) for mercury puts us into the range definitely shown to be toxic to people (1.0 ppm) if they eat more than about a pound a week. The warning signs are very clear. At a time when at least half the world already suffers from a serious protein shortage, and at a time when world population is doubling every thirty-five years, the continued pollution and destruction of the world's ocean fish must be regarded as criminal irresponsibility on the part of industrial peoples. In 1936, U.S. boats harvested 6,300,000 pounds of Pacific shrimp; by 1965 the total U.S. catch had dropped to a mere 10,000 pounds.[36] A fifth of all U.S. shell-fishing waters are officially closed by pollution;[37] almost all the rest *should* be closed because oysters, clams, and quahogs contain synthetic organic compounds (such as pesticides) and metals in dangerous quantities in almost every estuarine area of the nation.[38] Crabs in U.S. waters are now frequently found polluted by combinations of DDT, mercury, cadmium, lead, arsenic, manganese, zinc, and nickel. In other industrial countries, especially in Japan and Scandinavia, the destruction is even further advanced.

Most ominous is the effect of metals and other pollutants on phytoplankton, the tiny, floating plants which are known as the primary producers. As we saw in Chapter 1, through photosynthesis these organisms manufacture all the nutrients which make possible the ocean's food chain or food web. If it were not for the photosynthetic activity of the sea plants, all higher forms of sea life would starve to death. Lead, mercury, copper, silver, and numerous other highly toxic metals are now found concentrated in the bodies of phytoplankton and other microorganisms in the oceans of the world, especially in the critical estuarine waters.[39]

In biological systems, disruptions usually hit the highest predators hardest and earliest. As the primary producers succumb to pollution slowly, their numbers reducing gradually, the first and largest effects can be expected among the predatory populations of big fish and birds. If present trends continue, the higher trophic (feeding) levels—fish and birds—may disappear from the oceans completely. If this happens, people will be able to take food from the sea only by harvesting the primary producers themselves, plankton and seaweeds. If traditional human harvesting practices prevail, over-harvesting will occur (as has occurred with whales and numerous other valuable species) and this will lead to the final destruction of all forms of life in the oceans.[40]

These are still just speculations today, but they could quite easily become reality for our children. To turn these speculations into reality, all that has to happen is for enough of us to do nothing.

> Our technological society has committed a blunder familiar to us from the nineteenth century when the dominant industries of the day, especially lumbering and minerals, were successfully developed—by plundering the earth's natural resources. These industries provided cheap materials for constructing a new industrial society, for they accumulated a huge debt in destroyed and depleted resources, which had to be paid by later generations. The

conservation movement was created in the United States to control these greedy assaults on our resources. The same thing is happening today, but we are stealing from future generations not just their lumber or their coal, but the basic necessities of life: air, water, and soil. A new conservation movement is needed to preserve life itself.[41]

—BARRY COMMONER (1970)

11

The Decline of Affluence

Land, Metals, and Safe Sources of Energy as Diminishing Resources

"IT seems obvious," said an international science symposium recently, "that before the end of the century we must accomplish basic changes in our relations with ourselves and with nature."[1] It *seems* obvious, but how will it be done? Industrial modes of energy production, manufacturing and farming evidently must change because scientists from many diverse disciplines are reaching agreement that the earth simply cannot sustain current practices for very much longer. We are risking poisoning ourselves irreparably, and we are damaging (destabilizing) the oceans and other ecosystems which are essential to human life as we know it today. If we don't initiate fundamental changes very soon according to one eminent group of scientists, the "biosphere may not last for more than 100 years."[2]

A factor which very much complicates the spreading dangers of pollution is that we appear to be running out of the basic resources that make our current styles of industrial civilization possible: land, metals, and safe sources of energy.[3] In the United States—a nation gifted beyond all others in good, fertile soil—according to current predictions, all our tillable agricultural land will be in use within thirty-five years or so, about the time it will take our population to double. After that, there won't be any new land to open up.[4] For the rest of the world, the current prediction is even less optimistic: all tillable soil will be in use by 1985—just ten years from now. Af-

186

Table 11-1. World Average Rates of Increase for the Period 1955-1965 for Selected Aspects of Human Activity Related to Food Production.

Material	Percentage increase in constant dollars
Food	34
Tractors	63
Phosphate fertilizers	75
Nitrogen fertilizers	146
Pesticides	120

Table 11-1 adapted from Study of Critical Environmental Problems, **Man's Impact on the Global Environment** (Cambridge, Mass.: MIT Press, 1970), p. 118.

---

ter that, there won't be any new lands to open up.[5] "There has been an overwhelming excess of potentially arable land for all of history, and now, within 30 years (or about one population doubling time) there may be a sudden and serious shortage," according to a group of MIT systems analysts.[6] With the world population doubling so rapidly it is clear that pressure to use the land more intensively will continue rising. But as we have seen, it is intensive high-yield agricultural techniques which are severely and extensively damaging both soil and water now, as well as depleting world reserves of essential phosphorus. Table 11-1 shows the rates of increase in different sectors of industrial agriculture, 1951 to 1966; as is obvious, a small increase in the food supply requires much larger increases in the use of other materials (fertilizers, pesticides, and tractors). From this we can see that increasing pressure on the world's food supply will increase the stresses on the biosphere impressively in coming decades.[7] Table 11-2 and Figure 11-1 show why the pressure to increase food supply will be mounting as time passes: Human population has entered a phase of explosive growth.

Each year 70 million new people are added to world population; and each year between 10 million and 20 million people starve to death worldwide.[8] "Famine is no longer a threat but a reality in many areas," says Erik P. Eckholm of the federal Overseas Development Council.[9]

Even in the United States, we are experiencing tight food sup-

188

Table 11-2. Number of people added to world population each decade, 1900 to 1979.

Decade	Number of human beings added to world population
1900–1909	120 million
1910–1919	132 million
1920–1929	208 million
1930–1939	225 million
1940–1949	222 million
1950–1959	488 million
1960–1969	604 million
1970–1979 (projected)	848 million

Table 11-2 reproduced from Institute of Ecology, **Man in the Living Environment, A Report on Global Ecological Problems** (Madison, Wis.: University of Wisconsin Press, 1972), p. 17.

---

plies—allowing us to glimpse the fragility of the systems we depend on. The nation has been holding some 60 million acres of land out of production for the past several years. During 1974 with world stocks of grain down to less than one month's supply—perilously close to famine should crop failure occur—the United States removed all acreage restrictions to encourage farmers to produce more.[10]

The United States was expecting to put an additional 9,500,000 acres into production in 1974.[11] Despite this increase, weather conditions in the United States caused the 1974 corn crop to drop 12 percent compared to 1973. The late fall 1974 prediction was for a total harvest of feed grains in 1974 (corn, oats, barley, and grain sorghum) that would fall below the 1973 harvest by 17 percent.[12] Naturally this will affect the nation's meat supply, and the cost of food will rise again.

As the *Report of the Commission on Population Growth and The American Future* observed, "At a time when the federal government pays farmers to hold land out of production, it seems absurd to be looking forward to a scarcity of good agricultural land and rising food prices. Yet these are the prospects indicated by our analysis. . . ."[13]

New growth in cities, highways, reservoirs, and built-up areas

Figure   11-1 World Population

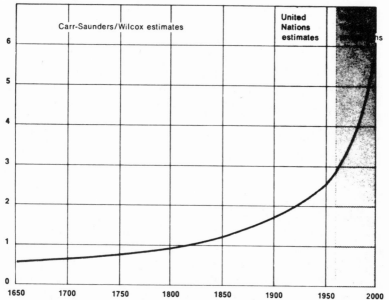

billions of people

World population since 1650 has been growing exponentially at an increasing rate. Estimated population in 1970 is already slightly higher than the
projection illustrated here (which was made in 1958). The present world
population growth rate is about 2.1 percent per year, corresponding to a
doubling time of 33 years.

SOURCE: Donald J. Bogue, Principles of Demography (New York: John Wiley and Sons,
1969).

Figure 11-1—Donella H. Meadows, Dennis L. Meadows, Jørgens Randers,
and William W. Behrens III, *The Limits to Growth* (New York: Universe,
1972), p. 33.

covers 1,200,000 new acres each year.[14] Much of this is prime
agricultural land. The national loss may not look significant now,
but much could look different just a generation from now. If present trends continue in California, for example, 50 percent of the
best acreage in the nation's leading agricultural state will be paved
over by the year 2020—just 45 years from now. Don't the nation's
children deserve to be left something?[15]

When we think of global population doubling in the next 35

years (by the year 2010), we ask whether the sea can continue to
serve as a vast source of food for the hungry billions. World population as of this writing is about 3.7 billion people. Humans in 1972
harvested more than 60 million metric tons of edible fish. But when
we look at estimates of the oceans' productive capacity, we find that
humans are probably only going to be able to harvest 100,000,000
tons yearly from the oceans, maximum. This means that we most
likely can't even double the present fish yields. In less than one
population-doubling time (thirty-five years) we will have reached
the limit of oceanic protein. This vast resource, which throughout
history has always been expandable, if things got tough enough will
not be amenable to further expansion.[16]

When we look at metals and other minerals, we see a different aspect of the same picture. Worldwide, the use of metals is increasing
at such a rate that every thirty years we consume approximately the
total amount of metals used in all previous time. "The entire metal
production of the globe before the start of World War II was about
equal to what has been consumed since."[17] In 1968 the National
Academy of Sciences–National Research Council drew together the
best available data on mineral supplies and published an authorita-
tive report entitled *Resources and Man*. This study made it clear that
contrary to popular belief, the world's mineral resources are not
some vast unlimited cornucopia of wealth. In fact, according to *Resources and Man*, right now we stand in imminent danger of depleting the world's known reserves of at least these key industrial minerals: silver, gold, platinum, tin, molybdenum, nickel, tungsten,
and mercury.[18] *Resources and Man* looked very hard at the widespread popular belief that abundant metals and other minerals will
one day be drawn from the oceans and from plain granite rock.
They rejected both these possibilities.[19] A 1972 study by an MIT
team is less optimistic. The MIT study estimates that even if we
multiply known mineral reserves by a factor of 5, we will still deplete the entire world supply of at least a dozen metals, plus petroleum and natural gas, in less than 100 years.[20] Even if current
estimates of global reserves are too low by a factor of 5, Table 11-3
shows, in less than 100 years we will exhaust world supplies of aluminum, copper, gold, lead, manganese, mercury, molybdenum,
natural gas, nickel, petroleum, silver, tin, tungsten, zinc, and the

Table 11-3. Essential industrial minerals in short supply worldwide.

Resource	Known global reserves	Years left at present growth-of-usage rates	Years left if we assume present growth-of-usage rates and five times known reserves
Aluminum	1.17×10^9 tons	31	55
Chromium	7.75×10^8 tons	95	154
Coal	5×10^{12} tons	111	150
Cobalt	4.8×10^9 tons	60	148
Copper	308×10^6 tons	36	48
Gold	353×10^6 troy ounces	11	29
Iron	1×10^{11} tons	93	173
Lead	91×10^6 tons	21	64
Manganese	8×10^8 tons	46	94
Mercury	3.34×10^6 flasks [of 76 pounds each]	13	41
Molybdenum	10.8×10^9 pounds	34	65
Natural gas	1.14×10^{15} cubic feet	22	49
Nickel	147×10^9 pounds	53	96
Petroleum	455×10^9 barrels	20	50
Platinum group	429×10^6 troy ounces	47	85
Silver	5.5×10^9	13	42
Tin	4.3×10^6 metric tons	15	61
Tungsten	2.9×10^9 pounds	28	72
Zinc	123×10^6 tons	18	50

Table 11-3 adapted from Donella H. Meadows, Dennis L. Meadows, Jørgen Randers, and William W. Behrens III, **The Limits to Growth** (New York: Universe Books, 1972), pp. 56-60.

---

platinum group (palladium, iridium, osmium, rhodium, and ruthenium).

Long before these metals and other minerals are gone, their

marketplace price will be continually rising. As *Resources and Man* said, "Unhappily, the widespread belief that technology is continually lowering unit costs while allowing us to work deposits of even lower grade is contradicted by the trends revealed in the copper industry."[21] The same report goes on to say:

> All postponements [including maximum possible recycling] allowed for, however, it is clear that exhaustion of deposits of currently commercial grade [minerals] is inevitable. Yet it is equally true that the increased costs of capital, labor, energy, and transportation for mining deposits of lower grade have in the past been largely offset by improved technology. The question is, can this continue? The answer remains to be given, but there are many signs that technology is not keeping pace with increasing costs of extraction in the United States.[22]

Resources and Man points to mercury as another example; during the past two decades the demand for mercury has doubled but in the same time period the unit cost of mercury has increased by more than 500 percent. The trend seems fairly clear; from now on, metals and other minerals will be costing us more and more, economically and environmentally.[23] Figure 11-2 illustrates that steeply rising energy requirements in U.S. mining industries are not matched by increased metals production.

It is in this context—the context of impending worldwide shortages in land and metals—that the so-called energy crisis in the United States takes on its most ominous dimensions.

The "energy crisis" has two dimensions: electricity and oil. We'll take electricity first. (We'll look at oil in the next chapter.)

Despite continual attempts by the energy industry to increase the demand for their product, in the early 1960s electric utility executives made a series of serious errors; utility executives actually thought that consumer demand for electric power would be dropping, not rising, throughout the 1960s. *Business Week* has termed this thinking "the conventional wisdom of the early 1960s." *Business Week* goes on to say:

> The conventional wisdom was given full expression in the Federal Power Commission's 1964 National Power Sur-

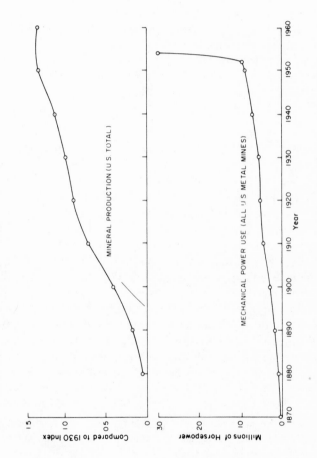

Figure 11-2. Steeply rising energy requirements in U.S. mining industries are not matched by increased metals production.

Figure 11-2 reproduced from Committee on Resources and Man [of the National Academy of Sciences—National Research Council], **Resources and Man** (San Francisco: W. H. Freeman, 1969), p. 122.

vey. The official line then was that most of the nation's
utilities were overburdened with severe plant over-
capacity. . . .
   Rarely has the conventional wisdom been so wide of
the mark. Seven years later utilities face not over-capacity
but overdemand.
   Utility planners persisted in their forecasting course
even after it began to look askew. So committed were they
to traditional theories of load growth that the deviations
were rationalized as weather aberrations. . . .[24]
   Around 1967 the planners suffered "future
shock. . . ."

   The kinds of errors that permeated the industry are instructive.
For example, the nation's largest electric company, Consolidated
Edison in New York, based its future demand predictions for air
conditioning only on construction of new buildings. Con Ed planners failed to realize that as new buildings put in air conditioning,
owners of old buildings would also install air conditioning to keep
their tenants from moving into cool new buildings. Thus starting in
November 1965[25] (when an overload put out the lights over an
80,000-square-mile area in the northeastern United States), Con
Ed was caught short and a decade later has still not recovered.
   Federal Power Commissioner Carl Bagge traced the source of
the industry's errors this way: "I submit that it was engendered by a
monstrous sense of intellectual and technological arrogance which
ignored not only the limitations of technology but, more impor-
tantly, the limitation of the vision of its high priests."[26]
   Fortune magazine, a bible of the business community, has described an even more serious error of judgment by utility executives:

   To meet the predictable demand for power, most utilities
   bet heavily and too hastily on nuclear plants, which are
   proving more difficult to build and more expensive than
   anticipated. . . .[27]

   As a result of these important mistakes by electric executives,
FPC Commissioner Carl Bagge, speaking before the American
Power Conference in April 1970, said, "I am obliged to make use of
this forum to . . . describe the situation confronting the electric

industry today as one which constitutes a 'national crisis.' "[28]

Circumstances left this key industry in the hands of a group of people whom *Fortune* magazine describes as "generally unimaginative men, grown complacent on private monopoly and regulated profits."[29] And now the nation is faced with a looming nuclear public health hazard of unprecedented magnitude. With demand for electricity still skyrocketing—and, in general, still being pushed by these "unimaginative men"—and with nuclear power not filling the demand fast enough, the electric utilities are now pushing harder than ever to get nuclear power plants operating as soon as possible. These same men are now rushing nuclear plants into use without testing one of their major safety devices and without confronting numerous other cogent criticisms of nuclear technology, many of which flow from inside the atomic industry itself.[30]

As of this writing nuclear power creates less than 6 percent of America's electrical energy, and this is the principal reason the industry is not yet critically dangerous—because nuclear power is used so little.

But the civilian nuclear reactor program, spearheaded by a national policy commitment to development of the breeder reactor,[31] is poised, ready to grow very rapidly in the next two decades. The rapid growth has already begun. In 1975 about fifty commercial reactors generated about 5 percent of the nation's electricity. By 1985 nuclear energy will be creating 25 percent of the nation's electricity, and by the year 2000, 50 percent or more of the nation's electricity will be produced by nuclear power.[32] By that time the technology will be impossible to turn off without causing a severe national economic depression.

Nuclear power plants present at least three distinct dangers to the public. First, there is a continual, normal, low-level escape of some 200 different radioactive materials into the environment.[33] This escape of low-level radiation may be controlled in the future, but to date control has remained inadequate.

Secondly, nuclear power plants put out quantities of hot, radioactive waste materials which must be contained indefinitely somewhere so they won't be dispersed into the environment.

Thirdly, nuclear power plants may undergo serious accidents which could result in release of massive amounts of radioactive materials into the environment.

Of these three problems, the first one can probably be solved.

However, control of low-level radiation may be expensive. Nuclear power is already more expensive than its competition (cleaned-up coal-burning plants for the short term and solar power at the end of the century). In the early 1960s, when the "stampede" to nuclear power first began, it was thought (by the AEC and by the companies that manufacture nuclear plants) that either a coal-burning plant or a nuclear plant could be built for about $110 per kilowatt. These estimates were wrong; a coal-burning plant ordered for delivery in 1975 cost $186 per kilowatt and a nuclear plant ordered for 1975 delivery cost $228 per kilowatt.[34] The repeated promise of cheap nuclear power has simply never come true.

The problem of low-level emissions from power plants is a serious one for this reason:

> It is universally agreed that the interaction of penetrating radiation and matter is such that in the case of living tissue, particularly in as complex an organism as man, the effects of exposure are overwhelmingly destructive. Furthermore, the amount of destruction appears to be proportional to the dose or quantity of radiant energy absorbed.[35]

This is a key point for understanding the hazard of nuclear power: *any* amount of radiation above natural background levels will do *some* damage to living things (especially people).

In 1969 Dr. Ernest Sternglass published frightening figures showing some 30,000 human cancer deaths per year attributable to pollution of the biosphere with radioactive materials. The AEC immediately assigned two of its most respected scientists, John Gofman and Arthur Tamplin, to assess Sternglass' data. Gofman and Tamplin found some of Sternglass' methods flawed, yet they themselves calculated that if the U.S. citizenry were subjected to the maximum allowable amount of radiation, 16,000 extra cancer deaths would show up each year in the U.S. population. Since this was clearly too high a price to pay for development of nuclear power plants, the AEC reluctantly responded by cutting the allowable dose of radiation by a factor of 10.[36] They have since proposed cutting it by another factor of 10.

The argument about the dangers of low-level radiation continues unabated. Sternglass insists his methods are *not* flawed. And there is increasing evidence that whether or not Sternglass' methods are rigorous enough, his results may very well be near the mark. A review by Allan Hoffman and David Inglis of Sternglass' latest book appeared in the *Bulletin of the Atomic Scientists* recently.[37] Hoffman and Inglis said,

> It is becoming increasingly apparent that radiation damage to human health, particularly in infants, may be several times more severe than was previously believed. . . . It appears that the bomb tests altogether may have caused several hundred thousand infant deaths in the United States along with thousands of maternal deaths, and fetal deaths possibly numbered in the millions.

The review goes on to quote Sternglass:

> "And for each infant dying in the first year of life, it is well known that there were perhaps three or four who would live with serious genetic defects, crippling congenital malformations, mental retardation, afflictions in many ways worse than death in early infancy." That could mean [conclude Hoffman and Inglis] about a million congenitally defective children in the United States as a result of nuclear testing,

Thus the problem of low-level radiation refuses to go away. On the contrary, low-level radiation continues to stir controversy among scientists. The specter of a plutonium economy has begun to dawn on the powers-that-be, and as we write, the nation is beginning to debate seriously whether we should go ahead with this technology.

By the end of this century nuclear reactors will be producing an estimated 80,000 kg (176,000 pounds) of plutonium each year. During the last quarter of this century the civilian nuclear reactor industry will create an estimated 10 million kilograms (22 million pounds) of plutonium.

Plutonium is a metal which does not occur naturally.[38] People created plutonium when they first fissioned the atom in the early 1940s as part of World War II's contribution to civilization.

Plutonium metal is pyrophoric; that is to say, it spontaneously

bursts into flame on contact with air. It then burns and gives off a
fume composed of very finely divided particles of plutonium. The
fume is just about the most toxic stuff that has ever been encountered. Plutonium has two lethal characteristics. First, it is chemically
a very strong poison. The lungs react to it by immediately initiating
a fibrotic response: lesions rapidly spread throughout the lungs,
suffocating the victim. A recent article described plutonium vividly:

> Cobra venom is nowhere near as toxic as plutonium sus-
> pended in an aerosol. You could hold an ingot of plutoni-
> um next to your heart or brain, fearing no consequences.
> But you can't breathe it. A thousandth of a gram of pluto-
> nium taken into the lungs as invisible specks of dust will
> kill anyone—a death from massive fibrosis of the lungs in
> a matter of hours, or at most a few days. Even a millionth
> of a gram is likely, eventually, to cause lung or bone can-
> cer. Plutonium that enters the blood stream follows the
> path of calcium. Settling in the bones it gives off short-
> range alpha particles, a form of radioactivity, and these
> effectively destroy the ability of bone marrow to produce
> white blood cells. [The victim develops the fatal disease
> leukemia, cancer of the blood cells.][39]

That, of course, is the second lethal characteristic of plutonium.
It is radioactive and causes lung cancer and cancer of the blood-
forming cells in the bones. Chemically, plutonium is very close to
calcium, and so far as the body's homeostatic mechanisms are concerned, plutonium acts like calcium. Plutonium is absorbed
through the lungs and stomach and stored in the bones. There its
low-level alpha radiation persists and causes cancer in laboratory
animals and humans. An ounce of plutonium efficiently distributed
could cause cancer in 30 million people.

The present health standards for plutonium allow a total body
burden of 0.06 micrograms. This is a very small amount of allowable poison, hundreds of times smaller than a single grain of table
salt. The nuclear reactor industry's present course, which will lead
to the creation of more than 20 million pounds of this material during this century, should make us all stop to think.

Although we recognize that it could never be distributed this
way, even one one-hundredth of one percent (0.01 percent) of this

plutonium, efficiently distributed, could kill more than 900 billion people—250 times the present global population.

Already losses are known to occur regularly in the nuclear fuel cycle. The losses are termed MUF—Materials Unaccounted For— and the MUF at a large fuel-fabricating facility (there are three in the United States so far) can amount to dozens of kilograms per year,[40] and this estimate may be quite low.

The routine loss of many kilograms of a poison which kills in less than microgram quantities is certainly an uninviting prospect, and the nation needs to debate the wisdom of proceeding on this course.

With a few thousand kilograms of plutonium in each reactor (breeder) there is no way anyone can keep track of the last gram. Gram by gram the accounting numbers gradually drift away from what one can find in stock, and one can never know if the problem is an accumulation of measurement errors, or a real theft, or a gradual loss as the material is repeatedly reworked. The oil industry used to consider a 2 percent inventory shortage allowable, and the nuclear industry will have its shortages. The implications of this fact must be faced.

As the civilian nuclear reactor industry presently functions, opportunities for error or misjudgment and opportunities for theft of nuclear materials abound. Here, for example, is a description of a typical shipment of fuel from a Japanese power reactor:

> In 1971 the Kansai Electric Power Company removed some fuel assemblies from its Mihama No. 1 reactor in Huki, Honshu, Japan. The radioactive fuel rods contained 50 kilograms of plutonium.
>
> In heavy casks the fuel assemblies were shipped to England. They went to Windscale—a reprocessing plant in Cumberland. Later that year, the fifty kilograms of separated plutonium, in oxide [powder] form, was shipped by BOAC to Kennedy International Airport. A courier rode along. At Kennedy the material was met by a man from Westinghouse and was loaded onto a truck (Forest Hills Transfer) that carried no other cargo. It was driven, on the New Jersey and Pennsylvania Turnpikes, to Cheswick, which is a few miles east of Pittsburgh. In Cheswick the 50 kilograms of Japanese plutonium has been fash-

ioned into pellets of mixed plutonium and uranium oxides, and placed inside seven hundred and fifty zirconi-
um-alloy rods. After the rods have been fitted into assem-
blies at another plant, in Columbia, South Carolina, they
will be shipped back to Japan.[41]

By the year 2000, according to the AEC forecasts, something more
than 1 million kilograms of plutonium will annually be traveling to
2,000 or 3,000 nuclear power plants in fifty-odd countries.

Power reactors are on line and making plutonium now in India,
Pakistan, East and West Germany, Japan and Spain. By 1980 thirty
nations will have reactors which are producing plutonium—which
can be processed readily into weapons-grade plutonium.

For that of course is another important aspect of the nuclear fuel
cycle. Plutonium is subject to diversion attempts by nations, small
groups, or even individuals. Once diverted, plutonium (or enriched uranium) could be used to fashion a simple, crude but devastating A-bomb.

Diverting plutonium from the fuel cycle actually might not be
difficult, given the way the cycle is operating today. Russell Wis-
chow, president of Nuclear Audit and Testing, Inc. (Washington,
D.C.), in 1973 said:

> I guess this is a dangerous statement, but I'm going to
> make it anyhow. If there were real intent to divert materi-
> al, you could get away with it. You can't be greedy. You
> have to work within the limits of measurement.
> What are you trying to do?—keep a bomb out of the
> hands of a country or a few grams out of the hands of a
> group? If you want a few grams of plutonium you can
> steal that almost anywhere in the country.[42]

Scientists argue about how easy it is to make an A-bomb. With
plutonium it's not all that difficult, according to Dr. Theodore Taylor, formerly of Los Alamos Scientific Laboratory, where he was
chief designer of the Davy Crockett atomic bomb (which weighs
less than 50 pounds) among others. Taylor has been speaking out
recently about the ease with which nuclear devices can be made. He
emphasizes that a nuclear bomb would really not be impossible for
a single skillful individual to manufacture. Taylor has said:

. . . .There is a level of simplicity that we have not talked about, because it goes over my [self-imposed and security-imposed] threshold to do so. A way to make a bomb. It is so simple that I just don't want to describe it. I will tell you this: Just to make a crude bomb with an unpredictable yield—but with a better than even chance of knocking this building down [the World Trade Center in Manhattan]—all that is needed is about a dozen kilos [25 pounds] of plutonium-oxide powder, high explosives (I don't want to say how much), and a few things that anyone could buy in a hardware store. [43]

If you take a grapefruit-sized mass of plutonium and compress it, squeeze it hard, using the force from a blast of high explosives, you stand a good chance of getting some kind of nuclear reaction. If you have done your homework diligently in the open literature on nuclear devices and if you have been careful, you may get a very large nuclear yield; if your work is crude, you may get only a fizzle yield. Even a fizzle yield could make a very destructive weapon, however. For example, Ted Taylor speculated:

A one kiloton bomb which could fit in a golf bag, or in the baggage compartment of a Volkswagen, exploded just outside the exclusion area during a State of the Union message would kill everyone inside the Capitol. It's hard for me to think of a higher-leverage target, at least in the United States. That bomb would destroy the heads of all branches of the United States government—all Supreme Court justices, the entire Cabinet, all legislators, and, for what it's worth, the Joint Chiefs of Staff. With the exception of anyone who happened to be sick in bed, it would kill the line of succession to the Presidency—all the way down to the bottom of the list. A fizzle-yield low-efficiency, basically lousy fission bomb could do this.

Henry D. Smythe, former U.S. ambassador (1961-1970) to the International Atomic Energy Agency, said in 1973,[44] "What I am concerned about internationally is power reactors in countries that have unstable governments. The Pakistan reactor, for example, builds up a stockpile of plutonium. Suppose there's a revo-

lution. A totally new and crazy government comes in and there's
plutonium just sitting there asking to be made into a bomb."

Since Smythe made his statement, India—one of Pakistan's bit-
terest enemies in recent years—has detonated a plutonium bomb
which it manufactured from reactor products.*

Even if a terrorist group did not create a nuclear detonation, plutonium could still be used very effectively for blackmail purposes. A
few grams of plutonium blasted into dust by a single stick of dynamite could contaminate—but contaminate for a quarter of a million years—a large piece of New York City or any other dense urban area. The costs of such an occurrence would be very large indeed. The blackmail potential of a criminal in possession of a few
grams of plutonium would be formidable and unprecedented. (We
sometimes hear it argued that blackmailers would have an easier
time threatening to put LSD in urban drinking water supplies. This
might be somewhat easier to do, but it would not have the uniquely
persistent lethality of plutonium dust. Just the contaminated real
estate costs of a plutonium attack on a major city would be extremely large, many billions of dollars.)

Another nuclear power problem, that of nuclear wastes, has
proved impossible to solve after more than twenty-five years of trying by America's best technologists. To date, the temporary solution has been to try to seal these wastes in steel containers inside
concrete shells buried in the earth. In the past twenty-five years the
U.S. nuclear industry has been holding wastes underground at the
following locations: Aiken, South Carolina; National Reactor Testing Station, Idaho; Richland, Washington; Oak Ridge, Tennessee:
Los Alamos, New Mexico; Albuquerque, New Mexico; Nevada
Test Site and Tonapah Range, Nevada; Portsmouth, Ohio; and
Paducah, Kentucky.[45] Altogether at these locations the nuclear industry has at least 80 million gallons of wastes stored away tem-
porarily, awaiting a final solution to the growing problem. High-

---

*The United States has so far firmly committed itself to provide enriched
uranium for thirty-three reactors overseas (France, two; Germany, two;
Italy, one; Japan, nineteen; Mexico, two; South Africa, two; Spain, one;
Taiwan, two; Thailand, one; Yugoslavia, one). The United States has
made a conditional commitment to supply enriched uranium to forty-five
additional reactors (Brazil, two; Greece, one; Iran, six; Korea, two; Neth-
erlands, one; Portugal, two; United Kingdom, two; France, six; Germany,
eight; Italy, one; Japan, eight; Spain, six).

level radioactive wastes are physically very hot (approximately 1,600° Fahrenheit, though they vary); they are also radioactive and will remain radioactive for up to 240,000 years, which is a very long time. These wastes are sunk in steel-lined concrete tanks which must *never* leak because leakage might permanently contaminate subsurface water supplies. Some of the containers have already been reported leaking.[46]

The Atomic Energy Commission (AEC) continues to look for a final resting place for the 80 million gallons of existing wastes, as well as for the quantities of additional wastes which will be created in the next few years. (By the year 2000, it has been estimated, nuclear power plants will be producing enough wastes to keep 3,000 six-ton trucks in transit at any given moment, hauling wastes to burial grounds.[47]

The AEC spent the past decade trying to convince the citizens of Lyons, Kansas, to allow the salt deposits deep under their town to become the final resting place for all of the nation's nuclear wastes. The citizens of Lyons (and Kansas state authorities) two years ago gave final rejection to the AEC's plan. Now the AEC has announced it wants to dump its wastes in southern New Mexico, near Carlsbad Caverns. As we write, a major fight is shaping up in New Mexico over the AEC proposal.[48]

Even if a spot can be found where the nation's nuclear wastes can be "safely" stored today, we will have to maintain a *perpetual* guard around the place to prevent people thousands of years hence from drilling into the ground and releasing deadly radioactivity by mistake. We must also regard the site in terms of earth crust movement, long-term erosion, glacial movement, intrusion of groundwater, and other geological changes which could contaminate vast regions with deadly radioactivity. Dr. Charles Hyder of the Southwest Research and Information Center (Albuquerque) points out that we must store these wastes for 240,000 years, and we're expecting the next ice age to descend on us in something like 10,000 years. "There's a serious moral problem, making life-or-death decisions for hundreds of generations of humans yet unborn," says Hyder.[49] Dr. Alvin Weinberg of the Oak Ridge National Laboratory has recently stressed the same aspect of nuclear power, saying, "We have created materials that man has never seen before, that remain toxic for times much longer than we have ever had experience with. We are forced, willy-nilly, to think on a time scale that exceeds

Pharoah's time scale 10- or even 100-fold. This is indeed unprecedented in human history."[50]

The third danger from nuclear power plants came to light in the early 1970s. The heart of a nuclear reactor is called the core. A reactor core contains approximately 200,000 pounds of nuclear fuel rods, as much radioactive material as is found in 1,000 Hiroshima-sized bombs. These radioactive materials slowly undergo a controlled nuclear reaction and produce heat. The heat is then carried off to make steam, which then turns a dynamo to make electricity. The reactor core is physically very hot and must be continually cooled to prevent it from melting. What happens if the cooling system fails? Scientists within the AEC itself, as well as independent investigators, have tried to predict what might happen if the "emergency core cooling system" were needed and failed. If a reactor's supply of cooling water were cut off and if the emergency cooling system failed, the temperature inside the core would rise to 2,000° Fahrenheit within 30 to 50 seconds; within 50 to 100 seconds the temperature would probably rise to 3,360°. Within 20 to 60 minutes the heavy, hot, radioactive puddle of melted core material would burn through the bottom of the reactor structure and begin burning its way down into the earth. This is known as the China Problem in the reactor trade because no one knows how far the dense radioactive mass might penetrate into the earth. As some wag suggested, it might come out in China.[51]

In the event of a melt-down accident, a cloud of radioactive material might be released, spreading long-lived lethal materials for several hundred miles downwind. The loss of life and property damage from such an accident could amount to hundreds of billions of dollars. For example, the Indian Point nuclear reactor is now operating at its location 26 miles from Times Square. A reactor is being constructed near Chicago's downtown area.

July 26, 1971, the Union of Concerned Scientists in Boston issued a report highly critical of the emergency core cooling system now in use on more than thirty operating reactors. The union said that the emergency system would "likely fail to prevent" a major catastrophe if an emergency developed. If the system did fail, radioactive materials might be dispersed into the environment in the pattern of the China Problem—or they might be dispersed for hundreds of miles across the surface of the earth as radioactive fallout in the pattern of a small nuclear bomb.[52]

In response to the Union of Concerned Scientists, the AEC admitted that the emergency core cooling system now in use has never been tested except through use of a mathematical computer model. According to the New York *Times,* AEC spokesmen have said the first actual full-scale tests on the emergency cooling system would get under way in 1974 but it now looks as though 1976 to 1978 is a more likely date. At that time the United States may have 100 reactors operating, all of them dependent on the untested safety backup system.[53] The AEC has not said what will happen to those reactors if the emergency system fails the full-scale tests. It seems to us unlikely that they would be removed from service, plunging perhaps 8 million U.S. citizens into darkness. In February, 1972, a coalition of sixty environmental organizations petitioned the AEC to stop licensing new reactors until after completion of the critical safety tests. The AEC has refused.[54] This refusal, of course, cannot inspire confidence in the AEC or in the hardware it claims is "safe." Reported in the open literature so far in the short history of nuclear power, there have been at least fifteen "incidents of concern to the public health" involving reactor malfunctions.[55] Looking at the record, physicist and science writer Ralph E. Lapp said in January, 1971, ". . . before the year 2000 we will probably have 500 nuclear reactors of one million kilowatt rating [each], and *it would appear a certainty we will have a serious nuclear accident*—and by that time population, if uncontrolled, will hem in our reactor sites." [56]

The threat of nuclear pollution from radioactive metals like uranium and plutonium has now become a peacetime specter of worldwide importance. Dr. Alvin M. Weinberg, director of the Oak Ridge (Tennessee) National Laboratory, has recently tried to point out "certain peculiar dangers of nuclear reactors which are small and unimportant today . . . [but which] may become overwhelmingly important when nuclear reactors become our dominant source of energy [.]"[57] Looking into the future, Dr. Weinberg sees a time when there may be up to 24,000 breeder reactors operating around the world (perhaps fifty years from now). Breeder reactors have special qualities which suit them to the modern situation; we are running out of uranium, which serves as fuel for current reactors, and the breeder will solve this problem. Using small amounts of uranium for fuel, the breeder reactor will create large quantities of plutonium; this plutonium can then be used as fuel for another reactor. Thus, the breeder reactor "breeds" new fuel, greatly extending the nation's useful nuclear fuel supplies.

We have already seen that the breeder reactor could also create several probably uncontrollable public health problems, among them the routine loss of small quantities of this extremely persistent, cumulative radioactive poison, and the spread of nuclear weapons.

The spread of nuclear power plants takes on particularly worrisome overtones when we place this development in the context of international events we can see developing in the next two or three decades. The world distribution of available resources is, today, grossly inequitable. The Western nations, plus the Soviet Union and Japan, now represent 27 percent of the world population, but together they consume 90 percent of the world's natural resources.[58] The United States alone, representing less than 6 percent of world population, takes more than 30 percent of the world's total resources each year. Such uneven distribution of resources has up to now been possible through the enormous military apparatus that the developed countries—especially the United States—maintain around the globe. In Latin America alone, the United States maintains an investment of $10 billion to assure the homeward flow of raw materials from our "good neighbors" to the south.[59] This is because "Economic chaos can result if foreign sources of supply are denied to a country that has allowed itself to become dependent on them," says *Resources and Man*.[60]

At present, all industrial nations except possibly the Soviet Union are net importers of most of the minerals and ores used by them. The dependence of the United States on foreign sources will almost certainly increase greatly during the next generation. . . . [S]ome of the metals most vital to the economic well-being of free-enterprise industrial nations are in areas of political instability or in Communist countries. Most of the known reserves of tungsten and antimony lie in such areas, as well as a large part of the world's manganese, nickel, chromium, and platinum. The present and near-future sources of manganese for North America are mostly in South America, Australia and Africa, the sources of tin in Southeast Asia, the sources of aluminum ore in various undeveloped tropical countries.[61]

Table 11-4 shows something about international relations and supplies of important metals.

For years the American people have kept hidden (though only from themselves) their original motives in the Vietnamese War. For years we told ourselves that we were fighting against "Communism" and for "freedom"—as if that very costly war were fought exclusively over ideas.

On April 7, 1954, President Eisenhower discussed the Indochina situation at a press conference; he said that the metals tin and tungsten were key raw materials which we needed to continue drawing out of Indochina. *U.S. News and World Report* expanded on the President's statement:

> One of the world's richest areas is open to the winner in Indo-China. That's behind the growing U.S. concern. Communists are fighting for the wealth of the Indies. Tin, rubber, rice, key strategic materials are what the war is really about. U.S. sees it as a place to hold—at any cost. . . . Southeast Asia's tungsten, tin and rubber, critically needed in the U.S. . . . Today, Southeast Asia's raw materials are still necessary to U.S. industry. . . . Southeast Asia's tungsten, iron, zinc, manganese, coal, antimony fit into Communist industrial aims.[62]

But the military might of the developed countries will surely not be able to maintain forever the system that has become known as neocolonialism and imperialism. As *Resources and Man* suggested in 1968, "Both rich and poor countries must be wary of the legacy of resentment that can come when a supplier nation finds that it has literally bartered its industrial potential for a mess of pottage."[63]

For years we could afford to risk the resentment that small, weak nations have always felt as they have watched the fruits of their heritage and their labor shipped overseas to the rich nations of the world. But now all that must change, for even the smallest nations in the world can soon be members of the nuclear club. Tiny Taiwan—the implacable island enemy of mainland China—has been given a nuclear power plant by Canada.[64] How long can it be before the Communist powers start giving nuclear reactors to the small angry nations in their camp? India's explosion in 1974 of a plutonium bomb tells the story of the future. A plutonium bomb manufactured from the waste products of a nuclear reactor does

Table 11-4. International relations and metals.

Resource	Countries or areas with highest known reserves (percent of world total shown inside parentheses)	Prime producers today (percent of world total shown inside parentheses)	Prime consumers (percent of world total shown inside parentheses)	U.S. consumption as a percentage of total world consumption
Aluminum	Australia (33) Guinea (20) Jamaica (10)	Jamaica (19) Surinam (12)	U.S. (42) USSR (12)	42
Chromium	Republic of South Africa (75)	USSR (30) Turkey (10)	(unknown)	19
Cobalt	Republic of Congo (31) Zambia (16)	Republic of Congo (51)	(unknown)	32
Copper	U.S. (28) Chile (19)	U.S. (20) USSR (15) Zambia (13)	U.S. (33) USSR (13) Japan (11)	33
Iron	USSR (33) South America (18) Canada (14)	USSR (25) U.S. (14)	U.S. (28) USSR (24) West Germany (7)	28
Manganese	South Africa (38) USSR (25)	USSR (34) Brazil (13) South Africa (13)	(unknown)	14
Molybdenum	U.S. (58) USSR (20)	U.S. (64) Canada (14)	(unknown)	40
Nickel	Cuba (25) New Caledonia (22) USSR (14) Canada (14)	Canada (42) New Caledonia (28) USSR (16)	(unknown)	38

Resource	Countries or areas with highest known reserves (percent of world total shown inside parentheses)	Prime producers today (percent of world total shown inside parentheses)	Prime consumers (percent of world total shown inside parentheses)	U.S. consumption as a percentage of total world consumption
Silver	Communist countries (36) U.S. (24)	Canada (20) Mexico (17) Peru (16)	U.S. (26) West Germany (11)	26
Tin	Thailand (33) Malaysia (14)	Malaysia (41) Bolivia (16) Thailand (13)	U.S. (24) Japan (14)	24
Tungsten	China (73)	China (25) USSR (19) U.S. (14)	(unknown)	22
Zinc	U.S. (27) Canada (20)	Canada (23) USSR (11) U.S. (8)	U.S. (26) Japan (13) USSR (11)	26

Table 11-4 adapted from Donella H. Meadows, Dennis L. Meadows, Jørgen Randers, and William W. Behrens III, **The Limits to Growth** (New York: Universe, 1972), pp. 56-60.

not require an expensive bomber or an intercontinental ballistic missile to deliver it. Such a weapon can be delivered on a fishing vessel or even conceivably shipped to its destination (disassembled) by commercial freight. In the face of such uncontrollable nuclear proliferation, the political and economic arrangements that have made possible our American affluence seem certain to change.[65] As Dr. Alvin Weinberg has recently said:

> We nuclear people have made a Faustian compact with society: we offer an almost unique possibility for a technologically abundant world for the oncoming billions, through our miraculous, inexhaustible energy source; but this energy source at the same time is tainted with potential side effects that, if uncontrolled, could spell disaster.

It is like a bad dream in which we see ourselves falling into the sea in a burning plane, and for some reason as a nation we refuse to wake up.[66]

> We are unanimously convinced that rapid, radical redressment of the present unbalanced and dangerously deteriorating world situation is the primary task facing humanity. . . .[I]n a world that fundamentally needs stability, [the] high plateaus of development [of the industrialized nations] can be justified or tolerated only if they serve not as springboards to reach even higher, but as staging areas from which to organize more equitable distribution of wealth and income world-wide.[67]
> —Executive Committee, the Club of Rome

12

Some of the Changes Necessary

"One guy. One guy and a fish."

BY now it is apparent to all informed people that we are creating each year many new threats to our biosphere, and to ourselves. Throughout the industrialized world, our flows of materials are very large and steeply rising. Each year hundreds of new chemicals and thousands of new processes go into use on a commercial scale, all part of a vast uncontrolled biological experiment we are carrying out on ourselves and on unsuspecting others.

We know that the higher predatory species (such as large seabirds and humans) are the ones most directly threatened by environmental contamination. We know that widespread damage is already occurring because when scientists learn how to look for damage, they find it—frequently in new, unsuspected, and disturbing places. Humans are fascinated by humans and so we know the most about pollution hazards to our own species. But we are shortsighted, and the most worrisome signs *should* be the subtle ones indicating that humans are systematically simplifying and destabilizing all of the earth's ecosystems to one degree or another. The long-term consequences of this simplification and destabilization cannot be predicted in detail but the general outlines of our fundamental predicament are crystal clear: continued destabilization of the earth's ecosystems must lead to ultimate collapse and disaster.

We are now in a position to understand why the newspapers and scientific journals are more and more frequently announcing the discovery of new threats to our biosphere and to ourselves. And we can understand why the pattern of recent discoveries is likely to

211

continue, at least into the near future, until something really serious occurs one way or the other. Look at the pattern that emerges if we just glance through the last two years or so of one professional journal, the American Chemical Society's weekly *Chemical Engineering News* (dates refer to when the various items appeared in the *News*):

●Vinyl chloride, a common synthetic organic compound has been discovered causing a rare type of liver cancer (angiosarcoma) among workers. Vinyl chloride reaches tens of millions of consumers in numerous ways each year. (April 1, 1974, p. 11)

●Deodorant sprays containing a metallic compound, zirconium chlorhydrate, are now suspected of causing lung tumors in humans. Some of the commonest underarm aerosol deodorants (Sure, Secret, Arrid Double X) contain the zirconium compound and perhaps as many as 100 million Americans have had their lungs exposed to zirconium by this method. (Ancient, so-called primitive humans seldom encountered zirconium at all.) As early as 1956 the manufacturers of zirconium-based creams and roll-on deodorants first learned that zirconium causes skin granulomas in some users. In lungs, granulomas lead to tumors. (December 9, 1974, p. 4)

●Hexachlorophene, a common additive in household cleansing bars (detergents intended for bathing humans), has been confirmed as a potent destroyer of brain cells in young children and most especially in prematurely-born infants. Infants whose skin was regularly washed with a common hexachlorophene-containing hospital cleansing agent (such as pHisoHex) would definitely risk severe brain damage and even death, according to research completed at the University of Washington School of Medicine. (January 20, 1975, p. 8) As we write, authorities are beginning to think about how to assess the costs associated with a long hospital history of hexachlorophene. Of forty premature infants that had been washed regularly with pHisoHex and which subsequently died, sixteen showed clear evidence of massive brain damage now known to be associated with hexachlorophene. Remember the ads for Dial soap, which contained hexachlorophene? How many babies were washed with it? Was the IQ of any of those babies lowered measurably? What would it cost to find out? Who should bear the burden of those costs?

●The common industrial monomer called vinylidene chloride

has been indicted as a cause of cancer in rats. (January 20, 1975, p. 8)

•The National Cancer Institute issued preliminary results indicating that another chlorinated hydrocarbon widely used in industry, trichloroethylene, causes cancer in mice. (May 5, 1975, p. 6; May 19, 1975, pp. 41-43)

•Toxaphene, a chlorinated hydrocarbon pesticide widely used as a replacement for banned DDT, has been found to cause a gross deformity of the spine in minnows raised in water containing only 55 parts per trillion (nanograms/liter). This is significant damage at very low levels of contamination. (April 21, 1975, p. 30)

•The National Institute of Occupational Safety and Health (NIOSH) has begun a search for information about the use of some 1,500 chemicals known to cause tumorigenic, neoplastic or carcinogenic responses in animals. These 1,500 chemicals were selected for special scrutiny from among the 12,000 toxic compounds listed in NIOSH's 1974 *Toxic Substances List.* (April 14, 1975, p. 6)

•Japanese researchers have reported excessive cancer rates among benzyl chloride workers. (March 31, 1975, p. 4)

•NIOSH has confirmed discovery of the spontaneous formation of a known carcinogen, bis-chloromethyl ether, from the reaction of two very common chemicals, formaldehyde and hydrochloric acid, in some textile plants. (March 31, 1975, p. 4)

•Chloroprene, a chemical compound which directly affects an estimated 2,500 workers and indirectly affects an unknown but larger number, is shown to have caused both cancers of the skin and lungs among Russian workers. (February 3, 1975, pp. 4-5)

•Plain inorganic arsenic has now been confirmed and reconfirmed as a cause of cancer in workers, and in the United States emergency measures are being taken to protect workers who encounter arsenic routinely. But workers' organizations are asking for stricter controls than the owners of industry are willing to install. (April 14, 1975, p. 5)

•Hair dyes used by 20 to 25 million Americans are now strongly suspected of being causes of cancer. Of 169 different hair dyes tested, 150 were found to be mutagenic when tested on microorganisms. The Cosmetic, Toiletry and Fragrance Association (a trade group) is vigorously disputing the evidence against the dyes. (March 24, 1975, p. 8)

•"Findings presented last week at a Conference on Occupational

Carcinogenesis, sponsored by the New York Academy of Sciences, indicate that the hazards of the work-place are spreading to families of the workers [and] into the general public." (March 31, 1975, p. 4)

•The New York Academy Conference heard reports of the discovery that there is a high incidence of cancer among people who administer anesthetics and there is a high incidence of birth defects among their children. There is an excessive leukemia rate reported among workers exposed to benzene, a very common industrial chemical.

The Academy Conference heard that cancer has now exceeded communicable and infectious diseases as a leading cause of death in industrial nations. The World Health Organization now estimates that 80 to 90 percent of all cancers are environmentally induced. Dr. Andrew Fairchild, Director of the National Institute of Occupational Safety and Health (NIOSH) estimates that cancer is costing the United States $35 billion a year now. This is an excessively costly disease situation, steadily worsening.

"So far government action in dealing with chemical carcinogens has been mainly reactive and fragmentary." (April 7, 1975, p. 18)

For four years, Congress has argued and debated toxic substances control legislation, but fruitlessly. This failure of Congress, more than any other, will be paid and painfully repaid by all of us but most especially by generations hence. Continued failure by Congress would be literally a crime against humanity.

•At the American Chemical Society's national meeting in September 1974, two researchers first announced evidence backing their theory that we are rapidly destroying the earth's ozone shield. (September 23, 1974, p. 27) It shook the meeting. As we saw in Chapter 1, the band of ozone (O_3) 12 to 25 miles high in the stratosphere protects earth from the sun's constant deadly bombardment of ultraviolet rays.

It has now been established by several studies that the ozone layer can indeed be destroyed by a class of compounds that industrialized humans have been releasing into the air for more than 20 years—chlorofluorocarbons (for example, the DuPont products called Freons). These gases are immensely useful in refrigeration systems, and they are widely used as propellants in household aerosol spray products such as deodorants, cosmetics, perfumes, insect sprays and automotive products like waxes and paints.

In other words, the pressurized gas from household aerosol products is slowly floating upward after we squirt our armpits, and many years later this persistent compound is reaching and destroying the stratosphere's ozone layer which was first established during primeval times. Once the chlorofluorocarbon gas reaches the ozone layer in the stratosphere, high-energy ultraviolet light decomposes it into chlorine atoms which then act as a catalyst that attacks ozone through a chain of reactions, ultimately turning ozone into oxygen. Oxygen doesn't filter out ultraviolet light the way ozone does, and the resulting increase in ultraviolet light on the surface of the planet is certain to cause an increase in human cancers, not to mention other important environmental ramifications.

Life couldn't crawl out of the ocean until the earth's ozone shield developed, and even 20 percent destruction of the earth's ozone shield would be extremely painful and costly just in human terms. The latest best estimates indicate that if our current rate of use of chlorofluorocarbons continues undiminished, we will destroy 16 percent of the earth's ozone shield within 25 years. This will result in 500,000 to 1,500,000 additional cases of skin cancer among Caucasian people per year worldwide; 100,000 to 300,000 additional cases of skin cancer would be expected among white people in the United States alone. A loss of 16 percent of the ozone shield could cause 20,000 to 60,000 deaths per year worldwide. The American Chemical Society's newsweekly has focused about a dozen articles on the ozone problem in the months since September, 1974. The National Academy of Sciences convened a special ad hoc panel October 16, 1974, which deliberated, then issued a statement saying the problem of ozone depletion by fluorocarbon aerosol spray products and refrigerants "is serious and should be given immediate attention by the Academy."[1] The Academy and numerous other groups have responded with research proposals and plans; the problem *is* serious. But let's get the basic lesson straight: Not only aerosol spray cans, but also supersonic transport airplanes (SSTs) and nuclear explosions can deplete the world's ozone shield. (*Chemical Engineering News,* April 21, 1975, pp. 21-23.) We may lose more than 16 percent of the ozone shield in less than 25 years if we don't guide *all* of our major technologies intelligently worldwide, and soon. The new threats to our biosphere are large and are real.

And they are rushing upon us suddenly from the horizon at an accelerating rate. *That* is future shock. And that is the future which we need to be preparing ourselves and our children to meet.

But why have the new threats to our biosphere only come to light since 1965? They have remained hidden for many reasons. First, they have traditionally been problems afflicting mainly workers and the poor. Only since World War II has the scale of human activities become so large that problems of the workplace and problems of the inner-city environment have now spread out and become problems of regional and global magnitude.

Second, the inherent delays in environmental systems have prevented us from appreciating environmental damage that has been going on for decades or centuries. Our failure to appreciate the delays in environmental systems has blinded us to the discovery of many fundamental environmental problems. Now we are learning, if slowly, to appreciate delays and the ways in which they have blinded us in the past.

Third, we've only begun to discover some of our most fundamental problems because of the nature of exponential growth. (A quantity exhibits exponential growth when it increases by a constant percentage of the whole in a constant time period. A savings account that grows by a steady 6 percent each year is growing exponentially.)

It is a characteristic of anything that's growing exponentially that it approaches a fixed limit with great suddenness. Take the world's arable land situation. We have a finite number of acres. Humans have about half of the world's arable acreage under cultivation today. To a simple observer, it still looks as if there's plenty of land left—twice as much as we now need. But human population is growing exponentially at a steady 2 percent per year; at this rate it takes only thirty-five years for the population to double. Today it may look as though there's plenty of land, but just one generation from now or sooner, the world will be up against absolute limits on arable land. The problem will "emerge" in just one twenty-five-year period and it will seem to come upon us with a rush (though it's a situation that's been building literally for centuries). As Paul Ehrlich and John Holdren have pointed out,

It should not be surprising that, when limits do appear,

they will appear suddenly. Such behavior is typical of exponential growth. If twenty doublings are possible before a limit is reached in an exponentially growing process (characterized by a fixed doubling time if the growth rate is constant), then the system will be less than half "loaded" for the first 19 doublings—or for 95% of the elapsed time between initiation of growth and exceeding the limit. Clearly, a long history of exponential growth does not imply a long future.[2]

Fourth, our current problems come upon us rapidly and unexpectedly because of the work of "indentured savants," to use the elegant euphemism coined by Dr. James Sullivan. By this, we mean professional people who are paid to lie, to mislead the public, to help corporations evade responsibility for their actions. This is the hired phalanx of shabby scientists, slick publicists, and venal lawyers who spend their lives delaying the removal of toxic materials from the environment and misinforming the public about the nature of the environmental crisis. They suppress data, they generate spurious data, they create false data, and they do their best to ignore important information that contradicts their own. They are visible in every public forum on environmental matters but of course their most important work goes on politically behind the scenes. They are the intellectual whores of industry, and they bear personal responsibility for much of the worst pollution and resource-waste that has been allowed to occur.

For more than 2 million years humans lived in balance with the biosphere. Now in just the past 200 years—and specifically more like just the past 30 years—humans have gotten badly out of balance with their sun-powered earth household. The earth-sun relationship that supports life is expected to continue for another 4 billion years, so the future existence of the biosphere would ordinarily be measured in terms of millions of millenia, essentially world without end.

But now thousands of informed scientists and lay observers are concluding on the basis of verifiable empirical evidence that the future existence of the biosphere— the living part of the earth—will be measured in decades. Recognition of this most fundamental predicament is now quite widespread. For example, a publication of New York's Chase Manhattan Bank recently stated the case succinctly: "The ecological balance is being dangerously disturbed in a manner that, if continued, threatens man's very survival. . . .The ecologists appear to have a valid case in terms of the long-term threat to the environment if postwar trends toward increasing pollution were to continue."[3] In 1971 the President's Council on Environmental Quality said we should not continue to allow the entire population or the entire environment to be used as a laboratory for discovering adverse health effects. There is no longer any valid reason for continued failure to develop and exercise reasonable controls over toxic substances in the environment.[4]

Among informed people there is, then, widespread agreement that pollution must be strictly and effectively controlled. The question is, how to achieve control? There are two possible basic strategies. The first, which is the one being tried by Congress now, is the regulatory-subsidy approach. This is the approach embodied in the Clean Air Act of 1970 and the Water Pollution Control Act Amendments of 1972.

Several recent studies of the benefits and costs of this approach have shown that the existing regulatory- subsidy programs for controlling air and water pollution are, and will continue to be, cumbersome, corruptible, arbitrary and capricious. The present pollution control strategy is also, and will continue to be, expensive, inefficient, and inequitable.

An alternative pollution-control strategy, called the pollution tax, has many advantages over the regulatory-subsidy approach. The pollution tax proposal has been best described by economists Allen Kneese and Charles Schultze.[5] Under the discharge tax plan, Congress would create a national tax on each pollutant—say 15 cents a pound for BOD (biological oxygen demand—a measure of a substance's pollution capacity) and say 30 cents a pound for sulfur. Each source of pollution would pay the discharge tax on each unit of pollution discharged into air, water or earth. Anyone dumping sulfurous wastes would simply have to pay the tax collector 30 cents for each pound of sulfur dumped anywhere. Since in the vast majority of processes, sulfur can be captured and contained for considerably less than 30 cents a pound, the polluter is strongly induced to control sulfur discharges. The size of the pollution tax can be varied to achieve as much pollution control as is desired.

Both the pollution tax and the regulatory-subsidy approach require monitoring programs to see what levels of toxic materials are being dumped by whom, but there the similarity ends. Here are some of the advantages of the pollution tax over the regulatory-subsidy approach. To begin with, the pollution tax provides a continuing incentive to develop and adopt new, less-polluting technology. On the other hand, the regulatory-subsidy approach gives a polluter no incentive to improve technology once a specified control level has been reached.

It is obvious that we need to continually improve our technology to tighten our levels of pollution control because the industrial system is growing worldwide at a rapid and accelerating rate. The industrial system grows 5 to 6 percent per year globally, doubling in total size every 12 to 14 years. Even if, by a massive effort, we could instantaneously cut our present pollution levels to 25 percent of what they are today, in two doubling times (totaling 24 to 28 years) the system would again be putting out as much pollution as it is now. Even allowing for an immediate factor-of-four decrease in pollution below today's levels, in one generation you're back where you started from, dangerously damaging the biosphere. Each doubling time thereafter, you're in severely deepening trouble. Therefore it is essential to build incentives into our pollution-control systems to encourage development and adoption of tighter and tighter control technologies. The pollution tax creates such incentives if it is vigorously applied.

A second advantage of the pollution tax is that it is more readily enforceable than the regulatory-subsidy approach. Both systems require monitoring of all discharges of toxic materials into the air, water and earth. But once the monitoring is completed, the two systems differ substantially thereafter. In the pollution tax case, the tax collector just has to collect the tax—something governments have known how to do for a long time. But the regulatory agency faces quite a different problem. Regulatory bureaucrats have to sit down, study all available options for the polluter, then dicker at length with the polluters' lawyers to reach agreement on what's the best option. The regulatory bureaucracy has been put in the position of having to be omniscient, of having to know the costs and benefits of each possible pollution-control strategy open to every polluter in the United States. Since there are 55,000 major polluters, of water alone, we can see that the bureaucracy needed to establish effective, efficient, equitable control technologies for each polluter must be very large indeed. As Kneese and Schultze have pointed out correctly, the regulatory-subsidy approach opens up a field day for lawyers, requires a large bureaucracy to give it any chance of success, incurs heavy costs, imposes severe economic impacts (especially on small firms) by ad hoc and capricious methods, and requires far-reaching intrusion of the government into decisions about the design of industrial processes. The pollution tax is clearly a better alternative.

A third advantage of the pollution tax is that, in any given region the tax encourages a pattern of pollution control among different firms and municipalities that tends to minimize the costs of control for the region as a whole. Several detailed studies have shown that the costs of the regulatory-subsidy approach can be as much as twice as high as the costs of the discharge tax approach, to achieve equal levels of pollution control. If you project costs of existing pollution-control plans into the future, it appears that the regulator-subsidy approach is perhaps going to cost Americans as much as $500 billion—half a trillion dollars—in the next twenty-five years. This is an excessive burden which will be felt by every citizen as a lowered rate of increase in standard of living. The saving of half this sum (by adoption of the pollution tax) would represent a saving of perhaps $10 billion each year for the remainder of this century.

In contrast to the regulatory-subsidy approach, the pollution tax system provides revenues. These could be used to finance regional-scale pollution-control agencies. For example, setting the discharge tax on BOD at 15 cents a pound would generate $2 to $3 billion annually. This would make a good start toward funding regional pollution-control agencies.

In addition, the present regulatory-subsidy approach ends up favoring particular technologies which may not be the most efficient or most effective ones to get the job done. The present air and water cleanup programs stress federal subsidies and tax breaks for waste treatment technologies. However, in any particular pollution case there may be better options than waste treatment—for example, the substituting of raw materials or the modification of basic production processes may be cheaper, easier and more effective. The pollution tax approach does not favor any particular technology, any particular method of environmental cleanup. It leaves polluters free to decide for themselves what's the best way to clean up, to avoid the high costs of the discharge tax.

Pollution taxes must be applied at levels which provide genuine incentives for control. And they will have to be applied gradually, to avoid severe economic disruption, but they should be applied starting now. Current water and air pollution control programs should be scrapped.

To establish a rational basis for the discharge tax system, the EPA will need to draw up a comprehensive list of toxic materials. The National Institute of Occupational Safety and Health (NIOSH) has already drawn up, by requirement of law, a continually-updated *Toxic Substances List*.[6] But the NIOSH list only includes chemicals known to present potential toxicity problems to humans experiencing relatively intense occupational exposure for limited periods of time.

The list will have to be extended to include all of the materials which, though they may appear harmless to human beings who encounter them in relatively high doses for limited periods in the workplace, may nevertheless at low, persistent doses cause serious and widespread damage to numerous living systems throughout the biosphere. For example, DDT is not very toxic to humans. People can drink an ounce or so with apparent impunity. Yet DDT has many subtle, complex, destructive effects on microscopic creatures upon whose health the whole biosphere ultimately depends. Moreover, DDT is an enzyme poison which damages the reproductive systems of numerous birds by preventing the deposition of calcium (a metal) in the structure of birds' eggs. The resulting thin egg shells crack when the parent bird sits on the nest in its normal way. The thin-eggshell problem is wiping out whole species of birds. Before this century is over, humans are expected to have exterminated up to 700 different species of birds and mammals, globally.[7] This is an impressive reminder of our terrible swift strength, which we are only beginning to dimly understand. Earth is covered by a vast, delicately balanced cyclical machine which is the living biosphere. We are tinkering with its parts, killing whole species at random and damaging others in important ways, without any real awareness of how the whole thing works.

Compare tampering with the biosphere to interfering with a vastly less complicated color television set. There is a possibility that, by splashing small amounts of hot solder at random into the circuits of a color television set, you could actually improve that particular TV set's performance. But the chances of bringing about an improvement by that method are very slim indeed. In the overwhelming majority of cases, splashing hot solder into a TV circuit causes problems.

As with a color TV set, so with the biosphere: random meddling without a detailed and comprehensive knowledge of how the system works in the vast majority of cases is liable to produce damage, to reduce the system's ability to work and survive. There was a long period of time when our meddling was so small-scale that the earth could endure our shortsightedness and our stupidities. But that time is gone forever. There are now too many of us and each of us can now unleash too much energy and release too many exotic materials.

The biosphere gives us numerous free services that we are accustomed to accepting without thought. The natural ability of ecosystems to decompose wastes, for example, is enormously useful to humans. The air conditioning systems of the atmosphere are another immensely valuable free benefit we take for granted.[8] It is when the limits of natural processes are exceeded that we see pollution, destruction and permanent damage. We must learn to perceive and measure the limits of natural systems, learn to accommodate our vast industrial machineries to the subtle and delicate (though large) natural systems of the earth. Effective, efficient, continuous management of our environmental resources *is* possible.[9] But to achieve it we must act rationally and decisively. And soon.

The pollution tax will not serve to control the most toxic and long-lived pollutants. The discharge of some substances must be flatly prohibited: the heavy metals which we've discussed would be prime candidates for a policy of complete containment.[10] Likewise, the discharge of certain persistent organic molecules, such as the aldrin-toxaphene group of pesticides, polychlorinated biphenyls, and the phenoxy herbicides (2,4,5-T and 2,4-D) among others, should be banned. The discharge of certain fine, persistent fibrous materials, such as asbestos and many varieties of fiber glass, should be prohibited by law and by regulation.[11] They are dangerous carcinogens.

Both a pollution tax and a ban on many toxic substances will bring net benefits to the whole society. However, we must go still further if we are to solve our fundamental problems. Somehow we must alter our personal patterns of possession and use. We must reduce the flow of materials needed to sustain our current wasteful lifestyles. To maintain the average person's style of life in America, we now extract from the earth (and then process) 40,000 pounds of new mineral materials for each person each year (see Figure 12-1) and, as Harrison Brown said in *Scientific American,* "This quantity seems certain to increase considerably in the years ahead."[12] The average American in 1970 required seven times the world average mineral usage per capita, and 100 times the average per capita energy used in the poor countries.[13] The U.S. population—less than 6 percent of the world total—uses one-third of total world energy production each year[14] and one-third of all the raw materials mined from the earth. By 1980 U.S. population will represent an even smaller percentage of world total, but we will be using perhaps 50 percent of the entire world raw-material supply.[15] All nations, but especially our own, must "encourage social and economic patterns that would satisfy the needs of a person while minimizing, rather than maximizing, the irreplaceable substances [the person] possesses and disperses."[16]

We can begin by eliminating artificial price advantages for virgin materials. As it is now, depletion allowances, freight-rate differentials, and other habitual incentives and penalties serve to boost the price of recycled materials and to give an unfair, unwarranted and destructive advantage to the sellers and users of freshly-mined, freshly-smelted products. These are destructive incentives which we must reverse. They served a nation of people who saw the need to conquer a virgin wilderness. But the conquering is done and the wilderness is a shambles of waste, mismanagement and abuse. The machine is ripping up the garden and the garden is the source of life itself. It is reasonable to turn those destructive incentives around.

Undoubtedly, as we move toward higher and higher levels of pollution control, prices are going to rise. It is important to point out that this will not be an inflationary price rise, but will instead reflect the embedding in the price system of costs that have heretofore been passed on (a) to particular sectors of the general public (chiefly workers and the inner city poor) and (b) to future generations through depletion of essential resources and through destruction of the biosphere.

Figure   12-1

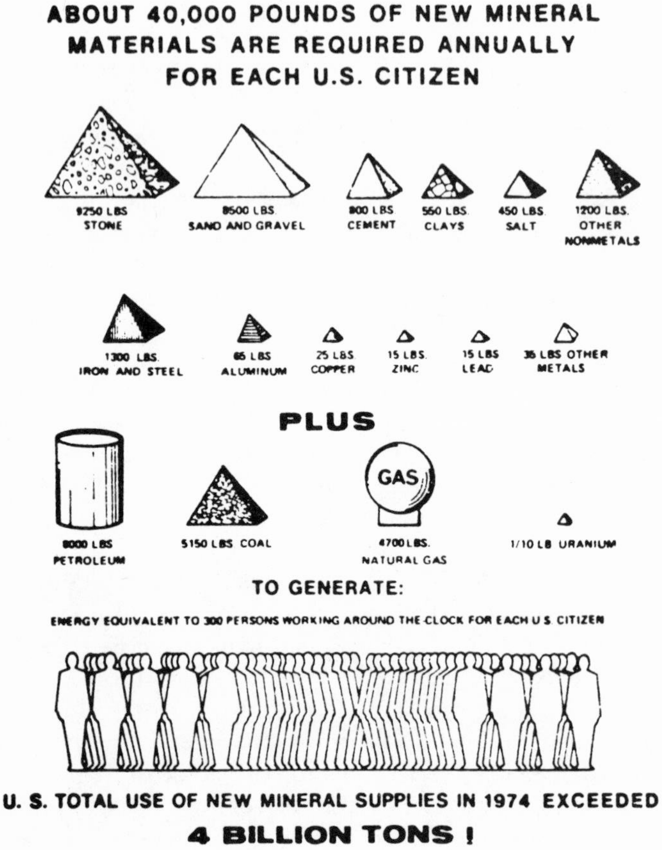

ABOUT 40,000 POUNDS OF NEW MINERAL MATERIALS ARE REQUIRED ANNUALLY FOR EACH U.S. CITIZEN

9250 LBS.
STONE

8500 LBS.
SAND AND GRAVEL

800 LBS.
CEMENT

560 LBS.
CLAYS

450 LBS.
SALT

1200 LBS.
OTHER
NONMETALS

1300 LBS.
IRON AND STEEL

65 LBS.
ALUMINUM

25 LBS.
COPPER

15 LBS.
ZINC

15 LBS.
LEAD

35 LBS OTHER
METALS

PLUS

8000 LBS
PETROLEUM

5150 LBS COAL

4700 LBS.
NATURAL GAS

1/10 LB URANIUM

TO GENERATE:

ENERGY EQUIVALENT TO 300 PERSONS WORKING AROUND THE CLOCK FOR EACH U.S. CITIZEN

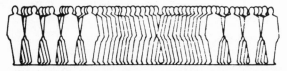

U. S. TOTAL USE OF NEW MINERAL SUPPLIES IN 1974 EXCEEDED
4 BILLION TONS !

from: Status of the Mineral Industries
        Bureau of Mines
        20 pp. Washington, DC: U.S. Dept. of the Interior, 1975

Paradoxically, as prices rise the quality of life will improve for everyone. The economic system will begin to reflect sound-ecology costs and quality-of-life costs, and through the mechanism of the price system the ingredients of prosperity will be distributed more widely and more equitably, and in balance with the biosphere.

A market economy can operate efficiently and effectively only if it begins with an optimum distribution of income. Two Massachusetts Institute of Technology economists, Lester Thurow and Robert Lucas, recently made this point very clearly in analyzing the U.S. economy. They said:

> All of the axioms that are used to praise market economies (capitalistic or socialistic) depend upon a fundamental condition. If a market economy starts with an optimum distribution of income, then a market economy will efficiently and equitably produce and distribute goods and services. Other conditions are necessary to insure that market economies really work, but the whole structure of justifications for market economies depends upon this initial condition. If the condition is not met, the most perfectly functioning market economy will be inefficient and unjust. It is simply starting out with the wrong distribution of economic voting power. . . .
>
> One of the main functions of government is to establish the right distribution of economic voting power. Not only must it establish such a distribution initially, it must continually re-establish such a distribution. Market economies will efficiently and equitably produce and distribute goods and services if they start with the optimum initial distribution of economic voting power, but market economies will not automatically generate such a distribution. Using tax and transfer policies, governments must be continually modifying market distribution of income.[17]

In redistributing income, Americans can look to three possible goals.

1. The ratio of the wealthiest 20 percent could be changed in relationship to the poorest 20 percent. Thus a definite idea of the desired wealth:poverty ratio would be one goal.

2. The distribution of incomes to minority groups could be

changed so that minorities (racial, sexual, ethnic, aged, etc.) become indistinguishable from the majority in their incomes.

3. Economic mobility could be another goal, to insure that a person's income is not determined by his or her parents' income.

Each of these goals involves value judgments. In fact, the political resolution of the value issues inherent in these three goals is considerably more difficult than achieving the desired income and wealth distribution after the three sets of goals have been clearly articulated and codified. The most efficient means for achieving any desired redistribution of income (and ultimately of wealth—since wealth is whatever produces net income) is simply taxation and direct transfer payments. The way to solve the value issues is by democratic processes—the modern equivalents of New England town meetings, elections, initiatives, and referendums.

As Table 12–2 shows, the ratio of the wealthiest quintile (top 20 percent) to the poorest quintile (bottom 20 percent) in the United States in 1970 was about 7.3 to 1. For the past 50 years, the annual per-capita income in the U.S. has grown at about 2 percent per year. In 1970 the average per-capita income was $3,130; by 1980 the per-capita income will have risen to about $3,815. If we had started in 1970 to change the distribution of the growth that could be expected to occur during the decade, we could have emerged in 1980 having eliminated poverty. The ratio of the income of the top 20 percent to the income of the bottom 20 percent by 1980 would have been 4.5 to 1, after redistribution had occurred. The bottom 20 percent would have experienced the greatest growth in income (measured as a percent of increase) and the top 20 percent would have experienced the smallest growth in income—but *everyone's* income would have continued to rise throughout the decade. No one would have lost anything that they had had in 1970. The top 20 percent would have lost some of the growth that would otherwise have occurred in their incomes during the decade, but they would still be considerably better off in 1980 than they had been in 1970— and there would have been no poverty left in our society. The *bottom* income for a family of four would be $6,500 per year.

Ultimately what is considered an optimum income distribution is a question of value; we have picked 4.5 to 1 somewhat arbitrarily. We believe that the question of redistribution of income and wealth is not one that can be settled except by a large-scale application of democratic processes.

The country is in the throes of a recession now and there's a possibility that it will turn into a full-scale depression, if economic controls are mismanaged. Despite such periodic downturns in business cycles, which have plagued American society since the inception of modern forms of corporate organization in the period 1850–1920, we believe it's fairly safe to say that per-capita income is going to continue to increase in the next couple of decades at about the same rate as in the recent past. That is to say, per-capita income in the United States is most likely going to increase at about 2 percent per year. If we look at the structure of income in the United States today, we can immediately see that expected growth in per-capita income is large enough to allow a redistribution of incomes without anyone losing anything they've got now.

Table 12–1 shows the projected growth of population and per-capita income over the next four decades. Table 12-2 uses these data to show how large the total economic-income pie will grow between 1970 and 1980, and how that growth might be easily redistributed to end all poverty in the United States.

The so-called War on Poverty begun under the Johnson administration was something of a farce, if not a fraud. The effects of its redistribution of income were very small or nonexistent. Recent

---

Table 12-1   Future population of the U.S. and projections of future per capita income.

Year	Population projection (in millions)	Per capita income projections	Total income projections (projected per capita income multiplied by projected population)
1970	203.2	$3130 *	6.36×10^{11}
1980	224.1	$3815	8.55×10^{11}
1990	246.6	$4650	1.15×10^{12}
2000	264.6	$5668	1.50×10^{12}

* Source of this number: James A. McCormick, "A Look at Poverty in New Mexico Through the 1970 Census," **New Mexico Business**, XXVI (June, 1973), p. 10.

Table 12-2 One possible way of redistributing income in the United States over the decade 1970 to 1980, to change the ratio between the Bottom 20 percent and the Top 20 percent from 1:7.3 to 1:4.5, effectively eliminating poverty.

Year	Bottom 20%	Second 20%	Middle 20%	Fourth 20%	Top 20%	Ratio of Bottom 20% to Top 20%
1970 (as income was actually distributed)	40.6 million people who receive $877 each (5.6% of the total)	40.6 million people who receive $1972 each (12.3% of the total)	40.6 million people who receive $2572 each (17.6% of the total)	40.6 million people who receive $3666 each (23.4% of the total)	40.6 million people who receive $6423 each (41% of the total)	1:7.3
1980 (how income might be distributed; assumes a 2% per year increase in per capita income)	44.8 million people who receive $1620 each (8.5% of the total)	44.8 million people who receive $2670 each (24% of the total)	44.8 million people who receive $3340 each (17.5% of the total)	44.8 million people who receive $4200 each (22% of the total)	44.8 million people who receive $7250 each (38% of the total)	1:4.5
% increase, 1970 to 1980	+85%	+35%	+21%	+15%	+13%	

Source: Lester C. Thurow and Robert E. B. Lucas, **The American Distribution of Income: A Structural Problem, A Study Prepared for the Use of the Joint Economic Committee, Congress of the United States** [March 17, 1972] (Washington, D.C.: U.S. Government Printing Office, 1972), p. 3.

efforts at tax reform have not been effective in redistributing income either. Whether you measure it before taxes or after taxes, the top 20 percent of the American people still earn about 7 times as much as the bottom 20 percent.

There are a variety of other income-redistribution mechanisms that could be used in conjunction with taxation and transfer payments. For example, by adjusting wages in various job categories, incomes can be redistributed. This is a potentially very important feature of any income-redistribution plan because the affluent may be willing to vote to allow poor people to earn higher wages whereas they might not be willing to vote to give the poor a larger handout. However, we should point out for the record that the direct handout is the most efficient means of redistributing income and wealth—it is the method that requires the smallest bureaucracy to administer, compared to all the more indirect (but perhaps politically more acceptable) means of income redistribution.

In any case, we should get rid of the idea once and for all that we can't end poverty. It would be easy to end poverty and relatively painless. The top income quintile might argue that the nation's economic rate of growth will be damaged by such a plan, but this is demonstrably not true. Their reasoning might be something like this: They have the largest incomes because they are the smartest (or their fathers or grandfathers were the smartest), the fittest, the best; therefore, society should leave its accumulated dollars (capital) in their trust, so that they in their wisdom can invest the money for society's maximum benefit. It's the same argument that keeps the private foundations in business—the old New England idea of the stewardship of wealth and virtue.

However, it has been demonstrated in other industrialized countries that the wisdom of investments (and therefore the economic growth rate) does not necessarily decline under a more equitable distribution than the 7:1 we have in the United States now. Japan, for example, has only a 5:1 ratio between its top 20 percent and its bottom 20 percent—and Japan has the highest rate of economic growth in the world.[18]

So there are no rational grounds that we know of for maintaining the present inequitable distribution of incomes and wealth. The pressing reason that makes redistribution of income and wealth *mandatory* and urgent is the need for effective environmental protection. The solution of numerous complex and threatening ecological problems is dependent upon eradication of poverty first in the United States and then worldwide.

There is growing awareness everywhere that the poverty-amid-plenty syndrome is truly unnecessary today. And there is some recognition that the changes necessary may be profound. For example, a recent publication of the Chase Manhattan Bank said, "No longer can traditional market forces be counted on to contribute to maximum social welfare."[19] As economist Allen Kneese phrased it recently, there is a "growing divergence between private ends and social ends."[20] The national Commission on Population Growth and the American Future (a commission created by the Congressional enactment of P.L. 91-213 and chaired by John D. Rockefeller, 3rd) recently concluded that

> The facts that we have cited describe a crisis for our society. They add up to a demographic recipe for more turmoil in our cities, more bitterness among our "have-nots" and greater divisiveness among all our people. What we have said here means that unless we address our major domestic social problems in the short run—beginning with racism and poverty—we will not be able to resolve fully the question of population growth.[21]

The Commission made it clear that they believe solving those problems—racism and poverty—domestically and on a global scale are matters of considerable weight and urgency. They said:

> The Commission has been deeply impressed by the unprecedented size and significance of the looming problems of resources and environment on a world scale. . . . We foresee potentially grave issues of clashing interests among nations and world regions, which could have very serious effects on the United States. . . . Our own future depends heavily on the evolution of a sensible international economic order, capable of dealing with natural resources and environmental conditions on a world scale.

All of the world's and all of the nation's problems are interrelated, and in a very real sense the solution of one problem (agricultural pollution, for example) requires the solution of all major problems.

Population control, the redirection of technology, the transition from open to closed resource cycles, the equitable distribution of opportunity and the ingredients of prosperity must all be accomplished if there is to be a future worth living.

Effective, efficient pollution control by the price system is only possible after redistribution of incomes. Then the costs of pollution-abatement or pollution-containment can be internalized into the price system through the pollution tax plan. Prices will rise, but we must not forget that it will be good for business to spend a significant fraction of our GNP on pollution-control technology. The expenditure will also reward consumers by the appearance in the marketplace of a wider variety of better-quality merchandise (food as well as other consumer goods). The rivers and beaches of the nation would once again be places fit for people to go swimming, the cities would be clean and would be improved in many other significant ways. Really cleaning up the environment would have all kinds of important secondary benefits such as longer lifetimes for many people, less disease, more interesting ways to spend leisure time, and better access to nature. Thought will further extend this list.

Ultimately, as we apply the pollution tax and develop regional pollution-control agencies, the nation may wish to firmly commit itself to a policy of "containment" as suggested by Theodore Taylor and Charles Humpstone in their provocative little book, *Restoration of the Earth.*[22] Whether the nation can adopt the policy of containment for all pollutants will depend upon the energy required to achieve 100 percent containment as advocated by Taylor and Humpstone. This question needs to be studied in detail. Clearly there are physical limits on the amount of energy-release the biosphere can stand. As we saw in the last chapter, changing the temperature of the globe just a few degrees could have very serious, far-reaching and irreversible effects. Ultimately, all of the fuels we use—including *all* the fossil fuels *and* geothermal power *and* nuclear fission *and* nuclear fusion—release energy which ends up as degraded heat discharged into the biosphere, the surface of the planet. This amounts to a net increase in energy injected into the biosphere, above and beyond the normal power input from the sun. A fairly simple mathematical relationship is observable here, indicating that a percentage change in net power input to the surface of the planet produces a percentage change in the planet's surface temperature one quarter as great. Average regional temperatures at different latitudes, which receive differing amounts of sunlight (the amount diminishing steadily from $10°$ north latitude, which is the planet's heat equator, to both poles) confirm this mathematical relationship.[23] In addition, the regional heat islands predictable from this mathematical relationship do in fact exist.[24]

We know reasonably well how much energy is being released by human activities worldwide. If we take the recently-observable rates of increase in energy use and simply project them into the future, we have to conclude that by the year 2050 (only seventy-five years from now), we will be adding enough waste heat to the biosphere to raise the temperature of the whole planet by something like 2.5 degrees Fahrenheit more. The ill effects of such global heating might include melting of the polar ice caps, among other undesirable consequences, as we saw in Chapter 11.[25]

The only energy source which will minimize the problem of overheating of the surface of the planet is solar energy. Thus we can see that ultimately the nation and the world are going to have to depend on solar power. Since it takes just about fifty years to shift an industrial nation's basic sources of energy, we know of no reason why the United States and other industrialized countries should not be planning and working aggressively now to have a largely solar-powered economy by 2025. For many low-temperature applications, technology for collecting and using solar energy is already available, as in hot-water heating and space heating.[26] In addition, solar-powered space-cooling technology is developing rapidly. Wind power is a form of solar energy we already know how to tap. Eventually silicon photovoltaic cells—or perhaps the recently-developed gallium arsenide cells—will produce electricity from sunlight with better than 10 percent efficiency and at reasonable cost.[27] There can be no doubt about this. Too much technology (defined as transferable knowledge) and hardware already exist, and are already in use, to deny the feasibility of heliovoltaic power sources on a major scale early in the twenty-first century. More than 90 percent of all of America's unmanned satellites have been, and still are, powered by small arrays (up to 20 kilowatt rated capacity) of silicon solar cells. One way to solve the problem of energy storage is by using electricity from silicon solar cells to electrolyze water into oxygen and hydrogen (probably using a multi-step series of catalytic reactions, though a relatively inefficient one-step electricity-to-hydrogen process has long been understood). Dilute hydrogen can be piped through existing pipelines and burned just about like natural gas and it is potentially even a clean source of fuel for internal combustion engines. Hydrogen can also be employed in low-temperature catalytic fuel cells to produce electricity quite efficiently.

People are afraid of hydrogen (H_2) because when the Hindenburg airship burned in 1937, the blaze was so tremendous and fearsome, many people jumped to their deaths rather than waiting to ride the burning ship safely to earth. Thus many people died and the event seized the national imagination and fixed us with a deep fear of hydrogen. However, hydrogen is much lighter than air and disperses rapidly upward while it is combusting. Although there *are* dangers associated with handling hydrogen, hydrogen fires and explosions are not nearly as frequent or as dangerous as many people think. Since 1956 hydrogen has been routinely shipped across the country in tanks, by rail and by truck, without an unduly bad safety record. Gasoline is dangerous to handle, too, yet we have managed to develop relatively safe means for handling it. Dilute hydrogen can be piped through existing pipelines (at some cost for additional compression stations) and it can be used by existing burners such as domestic stoves with only slight modification. The nation's gas distribution lines used to carry "town gas" which was 50 percent hydrogen before we shifted over to natural gas. The modification of burners required by that changeover was not an excessive economic burden. The changeover to a fossil-hydrogen, then solar-hydrogen economy could be accomplished with only minor inconvenience, though doubtless some expense at first.

The Mitre Corporation (a think tank) recently announced that they are installing a small (one kilowatt) heliovoltaic electrical system for their home offices using electricity from silicon solar cells to electrolyze water into hydrogen and oxygen which will be stored in tanks and converted back into electricity on demand by a fuel cell. Mitre Corporation's staff is not manufacturing any components—they're purchasing the whole system off the shelf. The technology all already exists. But Mitre is going to pay $130,000 for its one kilowatt system. If the system runs at 75 percent of design capacity for 50 years, producing 6,570 kilowatt-hours per year, the cost of the electricity will be 40 cents per kilowatt-hour—and Americans in

1975 are used to paying about 5 cents per kilowatt-hour. In addition, the system will certainly incur some maintenance costs in the fuel cell during the fifty years. On the other hand, the lifetime of the silicon solar cells—which so far are the really expensive part of the system—is expected to exceed centuries. Once they are created, silicon solar cells will convert sunlight energy directly into electricity with 5 to 15 percent efficiency seemingly forever. Discounting capital investments over the long term, we can justify a very large front-end investment in silicon solar cell technology research and development because it will pay permanent, significant benefits to all of humankind in the future, virtually forever. Silicon cells, or solar cells of some newer design, clearly hold the key to solving the planet's energy problems, and development of these technologies should be the nation's and the world's number one energy research and development goal. A solar-cell development program of Manhattan Project size could do the job in two decades or less.[28]

The development of silicon solar cells has numerous advantages over every other alternative. The cells are made of silicon, which is one of the most abundant elements on earth. Silicon is just sand. Silicon must be melted and then purified to a very high degree—an energy-intensive technology and a difficult technology to develop because of the high degree of purity required. (Molten silicon tends to dissolve everything it touches.) The ultra-purified silicon crystals are then purposely contaminated ("doped") with arsenic atoms and boron atoms. The final result is a thin rectangular or circular slab of dark blue glass, about as thick as three or four pieces of paper, each cell having an area of approximately one half square inch to one square inch. A thin wire is attached to each side of the cell. In the presence of sunlight, when the wires are attached to a load, electrons flow continuously through the wires giving pollution-free electricity. And the fuel is abundant, continuously renewed, and free.[29]

The silicon cells in use today convert sunlight directly into electricity with about 10 percent efficiency. Silicon cell efficiencies up to 26 percent are theoretically possible. With a gallium arsenide cell, over 20 percent conversion efficiency was recently demonstrated at two laboratories.[30] Large arrays of cells can capture very large quantities of sunlight energy and convert it directly into electricity. For example, the U.S. in 1970 used 2.1 trillion watts continuously (10,200 watts per person continuously). This was the total energy demand of the nation, including oil, coal, gas, uranium and electricity. By the year 2000 the total demand may double or even triple to a total of 6 trillion watts continuously.[31]

All of this energy could be provided in the form of electricity by solar cells operating even at 5 percent efficiency. The solar constant (the amount of sunlight falling steadily on the outside of the earth's atmosphere) equals 125.6 watts per square foot continuously. However, because of seasonal variations, daily variations (day-night cycle), and hourly changes in cloud cover, the amount of sunlight energy striking the surface of the United States averages 17 watts per square foot continuously. If this could be captured at only 10 percent efficiency, the total energy needs of the U.S. in the year 2000 could be met by solar collectors covering only 130,000 square miles (a square 360 miles on a side). This is 4 percent of the total U.S. land surface. We believe the 10 percent conversion efficiency is a reasonable estimate, but probably on the low side. Total heliovoltaic system conversion efficiencies, allowing for storage losses and other system losses, should easily reach 10 percent.

Naturally, nothing compels us to collect all of the nation's solar power in one location or even on land. Solar power invites decentralized systems and it invites floating offshore collector arrays. The sun shines for 2,000 hours during the year in northern Maine and 4,000 hours a year in southern Arizona—but the 17 watts per square foot continuously is a nationwide average figure. There is simply no doubt that heliovoltaic systems could supply all of the nation's and the world's energy needs into the foreseeable future.

There is no pollution associated with the operation of solar cells; there are no moving parts to wear out. So long as the cells are not covered by dust or snow, and so long as they are not broken by vandals, they will continue to provide electricity using the free fuel, sunlight.

The advantages of switching to a solar cell economy are obvious. The sun shines in abundance over much of the globe, especially over the most heavily inhabited parts, and it is fairly evenly distributed, so solar-powered machines can operate in both underdeveloped as well as developed countries, and in countries which have scant natural supplies of fossil or nuclear fuels. In addition, solar cell arrays can be designed to operate without any cooling water—thus making possible the generation of large quantities of electricity in the semi-arid and arid regions of the globe (approximately one third of the planet's land surface, including the south-western U.S.). These arid and semi-arid areas enjoy abundant sun-shine and relatively little cloud cover. Heliovoltaic power systems will do especially well in these areas, where cattle, horses, sheep, goats, and various wildlife can continue to graze the land beneath the solar collector panels.

Let's try to get the size of solar cell collector "farms" into perspec-tive. One of the biggest power plants in the United States is the Four Corners Power Plant near Farmington, New Mexico. It pro-duces 13 billion kilowatt hours per year. At a conversion efficiency of only 10 percent, a silicon cell solar collector measuring 5.6 miles by 5.6 miles (32 square miles) would put out as much electricity as the Four Corners Power Plant puts out annually. The Four Cor-ners Power Plant and associated strip mine and cooling pond take up more than 32 square miles today, not to mention the pollution and hydrocarbon resource destruction and water waste such a pow-er plant represents.

As has been pointed out in the case of the Mitre Corporation's heliovoltaic system, one obvious solution to the storage problem is electrolyzing water with solar-cell electricity to produce hydrogen which can be stored in tanks. Existing automobiles have already been modifed to run on a mixture of hydrogen and oxygen, as well as hydrogen and air.[32] There are engineering problems to be over-come, but a hydrogen-oxygen engine does not produce any nitro-gen oxides or any other pollutants except the low levels of trace ele-ment contamination associated with heating the lubrication prod-ucts needed by any metallic engine. The only principal exhaust product is what most people would consider pure water, eminently drinkable by today's standards. One car (a 1930 Ford pickup) rigged to run on a mixture of hydrogen and oxygen has as its ex-haust pipe a copper tube and at the end of it is a drinking cup cap-turing the exhaust. It's a tinkertoy idea right now, but it's basically not a bad approach. Perhaps the hydrogen will be stored in a vehi-cle and used in a fuel cell to produce electric locomotion. The first electric cars date back to the nineteenth century, after all. It is cer-tain that the technologies already exist to guide the nation through a petroleum, coal, hydrogen-solar conversion sequence to emerge in thirty to sixty years with a totally solar-based economy. Such an economy would use a variety of techniques with wind power, ocean temperature differential, heliothermal, bioconversion, heliovoltaic-hydrogen, and heliovoltaic-fuel cell systems supplementing the basic heliovoltaic collector panels.

Some otherwise knowledgeable people have disparaged the use of solar energy on a large scale because the collector mechanisms must be subsidized by fossil-fuel inputs. However, the manufacturers of silicon solar cells are talking now about powering their factories by solar cells; this would in effect create a "solar breeder" manufacturing facility. Solar cells creating electricity to create more solar cells to create more electricity. A net energy benefit is demonstrable. True, the manufacture of the first few megawatts of solar cell generating capacity must be subsidized by intensive fossil-fuel inputs; however, as our fossil fuel supplies dwindle in the next thirty years we would be wise to invest them heavily in solar breeder manufacturing facilities, because such facilities promise to extend the useful lifetime of our energy supplies indefinitely into the future. The energy systems of the future will all certainly *have to be* systems based on yielding net energy and useful work from the immense, dilute, interruptible flux of radiant power we call sunlight. The sun is the only way.

These are the technologies that the nation needs to be developing, not the nuclear power alternatives. As we have seen, particularly nuclear fission power is undesirable from numerous perspectives. Suffice it to say that nuclear power is fraught with hidden costs, and with not-so-hidden dangers. It is an extremely complicated "high" technology, orders of magnitude "higher" than solar cell technology (which is itself extremely advanced). The whole "peaceful atom" program begun by President Eisenhower in 1956 was developed as a national response of guilt over U.S. nuclear bombings. The "peaceful atom" fission-reactor program was ill-conceived from the start, doomed to failure in its search for a safe atomic source of cheap electric power for everyone. The pipedream has now permanently burst and the nation should simply stop throwing good money after bad with the civilian fission-reactor program, especially the nuclear breeder program. Fusion reactions should continue to be studied to see what they may yield in the way of important fundamental knowledge, but only on a limited experimental basis. Relative to the solar research budget, the fusion research budget is much larger than it needs to be, inflated by the Dr. Strangeloves who see doomsday weapons potential in the fusion research program.[33]

To begin with the solar research budget should immediately move up to the hundred-million-dollar annual level to match the fusion research budget. Anyone who says this level of funding couldn't be used effectively to develop solar technology is misinformed or dissembling.

During the next fifty years, the nation can change over to heliovoltaic, hydrogen and fuel-cell technologies, plus other alternative sources besides nuclear. In the meantime, coal combustion can be cleaned up significantly and the mining of coal can be made much safer and less environmentally destructive.[34] Coal can answer the short-term energy needs of the nation, insofar as those needs can be met at all, but no energy sources in sight now—fossil, nuclear or solar—can prevent the nation from having to tighten its energy belt in the coming decade.[35] The coming shortage is just the result of poor planning and of no planning. We are simply over-extended; we've built ourselves a lifestyle we can't maintain. It was a lifestyle based on future increases in imports of cheap petroleum and no such imports are available any longer. It was as poor a piece of planning as the nation has ever experienced. So we must look to energy conservation to get us out of the bind we're in for the short term, and in the medium term move ahead rapidly cleaning up coal and developing heliovoltaic, hydrogen and fuel-cell technologies. Eventually, we should be able to regain something like the standard of living we're in the process of losing, but it will take much ingenuity, much thought given to most efficient possible use of available energy. And it will take adequate funding of necessary research and development programs. We will not describe the many energy-conservation measures that we as individuals can take. They have been detailed by Dr. Al Fritsch of the Center for Science in the Public Interest in his readable book, *The Contrasumers* (New York: Praeger, 1974). These changes in lifestyle and technology are going to require a few new habits of thought among people everywhere. Here are some ideas we're going to have to become accustomed to:

Growth must soon slow down and eventually, in two to four decades, growth must stop completely. We mean growth of population, growth of industrial capital, and growth of per-capita energy and materials consumption. We are going to have to achieve a steady-state economy, an economy characterized by minimum throughput, maximum recycling of materials, and the most efficient possible use of energy. Moreover, the ingredients of moderate prosperity will have to be equitably distributed, worldwide and within regions and localities. The law of conservation of mass and the first and second laws of thermodynamics—physical laws of the universe which are immutable—absolutely determine that continued growth on a finite planet will inevitably lead to collapse of the world's ecosystems. So we know growth must end. The question remaining is: How close are we now to the limits to growth? The answers from chemists, physicists, and biologists are all converging on agreement: We are *very* close, only decades away from, major world ecosystem collapse, the permanent irreparable damaging of the biosphere on a massive scale.

We have just begun to think in terms of a no-growth economy and no one yet knows what such an economy will look like. Clearly feudal industrialization, mercantile industrialization, capitalism (in all its varieties), all existing forms of socialism and communism—indeed all existing industrial economies, whatever hybrid they represent—have failed to achieve a sustainable economic system fundamentally in dynamic equilibrium with the biosphere. It *is* possible for human communities to exist in dynamic equilibrium with the earth's self-maintaining, self-renewing cyclical ecosystems. Several billions of humans *can* be accommodated in moderate prosperity on the planet, but such a situation can be sustained only if it is done carefully and thoughtfully, through regional-scale planning and global-scale planning.[36]

The question of developing a new economic system is a fascinating one because it is so fundamentally important, and it should occupy many of the world's best minds during the coming decades. Traditional economics has never taken into consideration the finiteness of earth's resources, the finiteness of earth's regenerative and self-cleansing capacities.[37] Now these tangibles are coming to be measured and reckoned in our cost-benefit accounts. Howard Odum, at the University of Florida, and others are beginning to re-think all of economics in terms of energy flows.[38] For its part, the American Physical Society (the professional society for physicists) has been carefully defining the term "efficiency" and the Society is advocating that we re-think all of our transportation, manufacturing, processing, and housing technologies in light of a concept called "second law efficiency."[39] By this fundamental definition of efficiency, energy use in the United States is about 15 to 20 percent efficient. It should be and could be more like 80 percent to 95 percent efficient in these terms. The concept of "second law efficiency" is simple enough. Given a particular task to perform, physical laws determine how much energy will be required to do the necessary work. A particular technology is 100 percent efficient in terms of "second law efficiency" if it uses no more energy to perform the task than is actually necessary. If a particular process uses more energy than is actually necessary to carry out a task, that process is considered inefficient. Nationwide we use about five to six times as much energy as is theoretically necessary to do all the work we do. Our "second law efficiency" is about 15 to 20 percent. Obviously, careful re-thinking in the design of all our processes and many of our tools would yield tremendous increases in quality of life while allowing overall reduction in per-capita energy consumption. Endeavors and investments in this area will yield substantial rewards to society.

As we plan for a viable future, we must recognize the primary importance of the ecological principle, that the preservation of diversity is essential to the well-being of the planet's life-support systems.[40] Any unnecessary simplification of the earth's ecosystems is to be rigorously avoided. The destruction of diversity not only weakens the earth's life-support systems, but also robs future human generations of important genetic resources. Hundreds of medicines and thousands of other consumer goods now in use were originally developed from the unique characterisitics of particular plant and animal species. To destroy various forms of life is permanently to remove useful materials from the potential resource base of future generations. Such lack of foresight is unforgivable.

Among human communities a similar principle applies: Diversity is the basis of stability and survival. We must try to reverse the bleak modern trend so visible in industrial societies (especially in our own) and begin once again to cherish and foster cultural diversity. Instead of promoting sameness and superficially regimented standards of attitude and behavior, all our educational and socializing systems should recognize the basic biological fact that genuine health can be secured only by maintaining a diverse human and non-human environment. This principle must infuse not only our social, physical, and financial planning but our political processes and our private and public schools as well.

As we adjust to thinking in terms of maximizing diversity, maximizing second law efficiency, and minimizing growth, and as we develop sustainable economic systems fundamentally in dynamic equilibrium with the planet's non-human biogeochemical systems, we will need to re-emphasize several tendencies which have already emerged in American society.

(a) The process of environmental impact analysis written into the National Environmental Policy Act of 1969 should be strengthened and reaffirmed by its application to public and private developments at state and local levels as well as at the federal level. As a social process, environmental impact analysis opens up decision-making to a wider and better-informed public, and this makes the analysis of environmental changes a powerful tool for social betterment. It harnesses technical knowledge with political muscle in a democratic context. In one form or another, it is a tool that no complex technological society can do without, whether it's a democratic society or not. Any society which fails to institutionalize environmental impact analysis into its decision-making processes is bound to lose out in competition with other industrial societies in the long run (and even in the short run). As William Reilly pointed out in *The Use of Land*,[41] the key to successful implementation of the environmental impact analysis process is setting the right threshold for initiating the process. If you set the threshold too high and thus exclude too many small projects, the cumulative impact of many small projects will creep up on you and destroy the biosphere while you officially aren't looking.

On the other hand, if the threshold level is set too low, and you initiate the environmental impact analysis process too often, the costs of the process can become very great. Environmental impact analysis never fails to yield useful information, but the costs involved must be considered for they are social costs borne by the whole society.

(b) The process of technology assessment must be systematized and must spread throughout our society's decision-making processes. The creation of the federal Office of Technology Assessment and the creation by Byron Kennard and James Sullivan of the National Council for the Public Assessment of Technology (Washington, D.C.) are just beginnings of what must emerge as new analytic techniques are developed and are refined by study and use.

(c) We must develop genuinely useful Quality of Life indicators, and we must develop many kinds of measures of the status of ecosystems. Included in these developments must be new techniques

for environmental modeling, new development of biological indicators, new development of remote sensing tools and of other environmental monitoring systems. These developments must go forward rapidly—we are late in starting. The U.N.'s efforts in these di-
rections, and those of the Environmental Protection Agency and
the National Oceanographic and Atmospheric Administration
(NOAA), must be adequately financed by private foundations and
by governments.

(d) We should press ahead at maximum speed in developing
computer technology. It holds enormous promise for the solution
of human problems and the alleviation of human afflictions. The
low-cost hand-held electronic slide rule alone will revolutionize the
application of mathematics.The foreseeably developed low-cost,
hand-held programmable calculator will be a major new tool. The
immense information possibilities that are just now beginning to be
explored by the Commission for the Advancement of Public Interest Organizations (Washington, D.C.), and by others, should be
adequately funded and promoted. Their efforts, and the efforts of
private firms like Hewlett-Packard, Wang, and others should be expanded and promoted worldwide. Computers (and microfilm)
promise to give each of us, upon command, access to the full contents of the Library of Congress, indexed and retrieved paragraph
by paragraph on a TV tube via satellite or even printed on paper.
Our computer technologies help us think a million times faster
than we otherwise could. Information can replace travel and it can
replace work; information is power. Our national information ca-
pability is a fantastic strength, and it represents a source of immense energy driving our society. We should promote our computerized information technologies and use them to help get us out of
the multiple binds we're in. However, we should also be fully aware
that computerized information can present a serious threat to everyone's privacy, and that it represents other serious potential
threats to all industrialized societies. All power can be abused. Read
Herman Kahn and Anthony J. Wiener's *The Year 2000,* which is
not science fiction, and read *Privacy Journal,* a newsletter published
in Washington, D.C. Some of the prospects are chilling. Still, even
though the development of the microscope made possible germ
warfare and ultimately chemical warfare, we don't know anyone
who would oppose further refinement of the microscope, knowing
what we know today about its fantastic powers of discrimination.

Electronic computers represent an amorphous device that has been called a "macroscope," a machine that makes it possible to look at the universe of systems. The macroscope helps us see the whole immense forest in which we have so far only been narrowly observing individual trees. The macroscope is going to be seen one day as a more powerful tool than even the microscope; indeed it is seen by many in these terms today. Their good work should expand, and rapidly.[42]

(e) We need major new programs of *basic* biomedical research (as distinct from *applied* research) to establish accurately a fund of information about all our major pollutants in relation to environmental and human health. As we saw in the case of mercury and as we can expect in the case of many other pollutants, even if we stop all discharges into the environment today the problems will continue to get worse for years to come because of system delays. It is therefore urgent that we determine what levels of pollutants are dangerous in what ways to what life forms so that we can intelligently initiate controls. Present knowledge of these relationships is either extremely fragmentary or entirely nonexistent. We especially need to know the long-term, chronic effects of exposure to various chemicals so that we can evaluate the price we are already paying. At present we face an almost total absence of information about carcinogens, mutagens, and teratogens naturally occurring; without this sort of basic information, it is impossible accurately to assess the hazard potential that human activities are adding to the environment each year. What *is* known about teratogens (agents causing birth defects) is not comforting. A group of pregnant mothers were watched closely in a recent experiment. It was found that all of them—100 percent—were exposed to one or more known teratogenic agents during their pregnancies.[43] Half of them (49 percent) were exposed during the critical first trimester. Moreover, the researchers found that the average pregnant woman was exposed to multiple teratogens; the typical woman was exposed to an average of 6.3 teratogens, 5.4 of them being drugs. The categories of teratogens were: appetite suppressant pills, antiemetics (to combat vomiting), tranquilizers, analgesics (pain-killers), antibiotics, antihistamines, insecticides, acute illness, and radiation from X-rays. A total of 102 different known or suspected teratogenic agents were looked for in the study.

How many birth defects are "normal" and whether we've recently increased our rates of birth defects aren't accurately known. Since the thalidomide cases of the early 1960s we should have learned our lessons, but we haven't. In March 1972, more than a decade after the birth of 5,000 grossly defective children attributable to thalidomide, an eminent American teratologist was to write, "We need desperately and immediately a clinical surveillance program."[44] We urgently need to take a close look at a large sample of all births and follow them up to see if defects can be observed in the nursery. Only by large-scale clinical surveillance can we know if or how badly we are poisoning the next generations of Americans. But such programs are expensive and Congress has not yet felt enough heat and pressure to yield adequate funding.

Closely linked to the problem of teratogens is the even more difficult case of mutagens. Our genes transmit our human characteristics from one generation to the next with remarkable repeatability, given the complexity of what's going on. Our genes contain the chemical instructions that tell the next generation how to grow. Chemicals that interfere with the genes are called mutagens because they cause mutations.

A gene damaged by a mutagen now may not show up immediately in an offspring. The damaged gene may be carried forward for several generations before it manifests itself as a bad liver, or a weak back, or an arterial system subject to clogging, or cancer of the pancreas, or some other illness or debility. Mutagens can affect any and all of the body's chemical systems, causing weak eyes or decreased resistance to germs or an excessive requirement for a particular nutrient (vitamin or mineral). About 1,500 different human diseases are thought to be genetic in origin.[45] Dr. Samuel Epstein has stated that we need a massive effort to monitor human populations to observe the mutation rate, to determine whether or how we are changing the rate.[46]

Without direct observation of humans, the job is very difficult. Even the best, most expensive laboratory experiments rely on exposing test animals to chemicals to try to produce damage. In the case of the extreme poison beryllium metal, the debilitating lung disease couldn't be produced in laboratory animals. This fact delayed for a long time the recognition that beryllium really was causing human disease.[47] Several hundred people paid for that delay with their lives. In the case of thalidomide, it turned out to be much more teratogenic in people than it was in the animals on which it was tested prior to marketing. It turned out that human beings were 60 times more sensitive to thalidomide than were mice, 100 times more sensitive than rats, and 200 times more sensitive than dogs.[48]

A few scientists within the FDA and the President's Council on Environmental Quality and a handful of Senators and Congresspeople appear to have a good grasp of these problems, but almost everywhere else in the federal administration we observe a distinctly parochial and uninformed attitude. One eminent physician, Dr. Robert Brent, recently summarized the situation for a Congressional committee:

> Basic biomedical research in this country is in serious jeopardy. There is a despair that is apparently not evident to the Congress and the public, and little is being done to deal with the crisis in basic research because they see the turmoil of their faculties and the shrinking support for training opportunities in research.[49]

Dr. Brent went on to make a crucial point about the way basic scientific advances take place. He said:

> How did this new climate develop? The public was deluded and disappointed by the implied promises of research programs that purported to solve certain medical problems. . . . Applied research programs were instituted to solve clinical problems before enough basic information was available to initiate a rational approach. Too many of our applied programs were doomed to failure.
>   . . . Would we have been able to develop a polio vaccine if Harrison in 1902 had not developed the techniques for growing frog's cells outside the frog's body? The development of these tissue culture techniques permitted Enders to grow polio virus outside the body. It was then easy for Salk and Sabin to do their important applied research. New concepts and new ideas that are important for mankind lie beyond the horizon for most scientists. It is basic science research that expands those horizons.

The American electorate must be educated to the continuing need for *basic* scientific research. Research priorities guided by the typical politician's or businessman's goal (short-term dollar profit) will lead the nation down a very dangerous road. We desperately need tremendous sums of money for research to understand (and later to solve) the problems that threaten the continued existence of our nation and of the human species itself, but the American people will not support such research unless they understand something about science itself and how it makes advances. More scientists and fewer businessmen must guide the establishment of the nation's research priorities; otherwise the money will probably be largely wasted, as huge sums are being wasted today. It is no doubt politically expedient to announce that one is about to "solve" our problems by getting "practical" about research. But when the solutions are not forthcoming, the American people may turn against all of science, as they seem to have been doing lately, with unfortunate long-term consequences for us all.

Science in America has got a bad name perhaps because so many scientists have sold their services to the military. American science has developed a "death wish," according to Lewis S. Feuer, and the end of all science may result.[50] This would be disastrous for the future of all people. Despite its close association with the most destructive forces in our society, science must also be recognized as our only reliable way of achieving widespread agreement about the nature of reality. The American philosopher-scientist Charles Sanders Peirce made this clear in a brilliant essay published almost a century ago, in 1877.[51] Peirce saw correctly that the method of science is the only way of knowing about reality which allows people of diverse cultures to come to almost universal agreement. Rough agreement about the nature of reality is required before people can act effectively together to achieve common goals, such as survival of the human species. Science, more than any other way of arriving at knowledge, has brought widespread cross-cultural agreement (as among astrophysicists, chemists, and soil scientists) about the nature of reality. No method of acquiring knowledge, except the scientific method, has achieved such widespread enduring agreement.

We are not advocating that scientists take over decision-making processes. As Barry Commoner argued cogently in his first book, *Science and Survival* (1966), scientists are no more qualified to make ethical judgments than anybody else, and political questions are always questions of value, of ethical choice. But scientists *do* play a special role in the modern world, whether they like to recognize and fulfill this role or not. The atomic bomb changed the role of scientists in society irrevocably. Commoner argues, and the Center for Science in the Public Interest (Washington, D.C.) argues, that scientists must get involved in the day-to-day decision-making processes by providing relevant, valid and reliable information at key points in the political process. Scientists must get involved in letting people know what facts are and what *the* facts are.

Now we must turn to the important problem of changing corporate behavior. Although multinational corporations can in particular instances be effective organizations for promoting, for example, the spread of useful technologies, these same corporations also have some of the dangerous characteristics of immense sharks. This has already been demonstrated amply and will not be argued in detail here. We simply make the point to lay the background for the next set of recommendations.

Corporate behavior is going to have to change, whether the size of corporations changes or not. The shark is going to have to permanently lose many of its teeth, because in its modern immensity it threatens to chew up too many people and too much of the world's nonrenewable resources. It has developed huge, inexorable appetites. We must try to get into perspective the sizes of the corporations we are talking about. Take just the oil companies. Table 12-3 shows how large the U.S. oil companies are, comparing their gross annual sales with the gross national products of countries. As is plain in Table 12-3, Standard Oil of New Jersey (now calling itself Exxon), is larger than Denmark, larger than Austria, larger than Yugoslavia, larger than Indonesia, larger than Bulgaria, larger than Norway, larger than Hungary, larger than the Philippines, larger than Finland, larger than Iran, larger than Venezuela, larger than Greece, larger than Turkey, larger than South Korea, larger than Chile. The second-largest petroleum giant, Mobil Oil, is larger than Colombia, larger than Egypt, larger than Thailand. Texaco is larger than Portugal, larger than New Zealand, larger than Peru, larger than Nigeria, larger than Taiwan. Gulf Oil is larger than Algeria, larger than Ireland, larger than Malaysia. Standard Oil of Indiana is larger than North Korea, larger than Morocco, larger than South Vietnam, larger than Libya, larger than Saudi Arabia.

Table   12-3. Nations and multinational corporations compared in size.
One way to show the size of today's large multinational corporations is
to compare their gross annual sales with the gross national products of
countries.   This table uses 1970 figures for all except the centrally
planned economies (excluding China) and General Motors Corp., for
which 1969 figures were used.   The amounts are shown in billions of
dollars.

1	United States	$974.10	37	Philippines	10.23
2	Soviet Union	504.70	38	Finland	10.20
3	Japan	197.18	39	Iran	10.18
4	West Germany	186.35	40	Venezuela	9.58
5	France	147.53	41	Greece	9.54
6	Britain	121.02	42	Turkey	9.04
7	Italy	93.19	43	GENERAL ELECTRIC	8.73
8	China	82.50	44	South Korea	8.21
9	Canada	80.38	45	IBM	7.50
10	India	52.92	46	Chile	7.39
11	Poland	42.32	47	MOBIL OIL	7.26
12	East Germany	37.61	48	CHRYSLER	7.00
13	Australia	36.10	49	UNILEVER	6.88
14	Brazil	34.60	50	Colombia	6.61
15	Mexico	33.18	51	Egypt	6.58
16	Sweden	32.58	52	Thailand	6.51
17	Spain	32.26	53	ITT	6.36
18	Netherlands	31.25	54	TEXACO	6.35
19	Czechoslovakia	28.84	55	Portugal	6.22
20	Romania	28.01	56	New Zealand	6.08
21	Belgium	25.70	57	Peru	5.92
22	Argentina	25.42	58	WESTERN ELECTRIC	5.86
23	GENERAL MOTORS	24.30	59	Nigeria	5.80
24	Switzerland	20.48	60	Taiwan	5.46
25	Pakistan	17.50	61	GULF OIL	5.40
26	South Africa	16.69	62	U.S. STEEL	4.81
27	STANDARD OIL (N.J.)	16.55	63	Cuba	4.80
28	Denmark	15.57	64	Israel	4.39
29	FORD MOTOR	14.98	65	VOLKSWAGENWERK	4.31
30	Austria	14.31	66	WESTINGHOUSE ELEC.	4.31
31	Yugoslavia	14.02	67	STANDARD OIL (Calif.)	4.19
32	Indonesia	12.60	68	Algeria	4.19
33	Bulgaria	11.82	69	PHILIPS ELECTRIC	4.16
34	Norway	11.39	70	Ireland	4.10
35	Hungary	11.33	71	BRITISH PETROLEUM	4.06
36	ROYAL DUTCH SHELL	10.80	72	Malaysia	3.84

A multinational corporation has been defined as one with annual sales above $100 million, with operations in at least six countries, and with overseas assets accounting for at least 20 percent of its total assets. There are 4,000 companies in the world which qualify.

Corporate control must be asserted by government. It *is* possible. Here are two suggestions taken from Ralph Nader: First, all corporations should be chartered by the federal government.[52] At present, the individual states charter corporations. The states compete with each other to get corporations to locate within their jurisdiction. As a result, the states cut each other's throats by setting taxes as low as possible and doing as little regulating as possible. The state that promises the fewest obstacles ends up as home base for the biggest number of wealthy corporations. (Delaware has been winning this competition in recent decades.)

If the federal government chartered corporations, then the federal government could place strict controls on corporate behavior. This is clearly established in existing law. Any corporation that didn't operate in the public interest could actually have its charter revoked. This would be a real lever to force reform of corporate behavior.

A second important change which would give more control over the multinationals would be to reinterpret the legal status of corporations. It was in the early 1890s that the U.S. Supreme Court first interpreted the Constitution to give corporations the same protections under the law that individuals enjoy. The U.S. Constitution was drafted during the eighteenth century when there was extreme suspicion of oppressive government. As a result, the Constitution gives individuals many good protections against arbitrary authority. But the extension of due process and individual rights to corporations changed the picture significantly. It is one thing to give a single person the right to due process and it is quite another to extend those rights to multibillion-dollar organizations. If they choose to, the Big Eight oil firms alone can overwhelm the federal government's relatively small battery of lawyers on any important question. The same is true of every other oligopolistic and essential sector of modern industrial America. Regulatory officials know this; they consequently avoid direct confrontation with the oil firms (or other large corporations such as the automobile giants). The government bargains with the corporations, settles out of court, ac-

cepts compromise terms and cosmetic reforms. The result is the
world in which we live today—a world in which all the people of the
United States have lost control over the necessities of life, such as
their sources of fuel, healthy clean air, and water free of carcinogens. To alleviate this situation, the Supreme Court or the Congress could remove from corporations some of the protections
of due process. Corporations could become a special class of
"individuals"—which they in fact already are, since they are so
enormous.

Another important element in a program to control corporations
more closely would be to give individuals greatly-expanded rights
to sue under a broadened conception of nuisance doctrine. For
more specific ideas, see the book *Defending the Environment,* by Joseph Sax, in which the author describes ways of making the legal system a more responsive and effective tool for environmental
protection.[53]

Finally, making corporate officials personally answerable for corporate behavior would go a long way toward bringing private corporate decisions more closely into line with the broad, stated policy
goals of the nation. If more corporate officials had to go to jail for
their crimes against the environment, against workers, and against
the "general public," the silent but deadly violence of pollution and
resource-waste might be curbed more effectively than we're man-
aging today. Making corporate officials personally answerable for
corporate crimes would have the added social benefit of immensely
accelerating the rate of penal reform throughout the country.
Once a few corporate decision-makers had tried a modern peniten-
tiary on for size, we might see a lot of changes in a hurry.

Some will argue that an "incentives" approach would be more
persuasive to corporate leaders. However, the incentives approach
has already been tried—we have made the nation's corporate owners and leaders rich and powerful beyond the wildest dreams of
former kings; we have rewarded them generously and given them
enormous decision-making powers over us. In turn we have been
repaid with the haphazard development of an enormously inefficient, wasteful, inequitable, and non-sustainable industrial system
in which 20 percent of the system's participants are more or less
permanently prohibited from enjoying the fruits of the society's
combined intelligence, knowledge, energies and labors. Something
like 25 million Americans average less than 30 cents per person per

day for food, 30 cents per person per day for housing, and 30 cents per person per day for the remainder of their necessities such as educaton, transportation, medical needs, dental care, and recreation. Such poverty amid plenty represents a failure of corporate and governmental imagination of breathtaking magnitude.

In addition, according to a study by the U.S. Senate's Antitrust and Monopoly Subcommittee, as reported by Senator Philip Hart of Michigan, in 1969 American business took from consumers somewhere between $174 billion and $231 billion that purchased not one cent of product value. (It's the old water-in-the-frozen-orange-juice trick writ large.) Since it has been reported by law enforcement officials that businesses lost less than $16 billion as the result of non-organized crime in 1971, the statement seems to be correct that, in 1969 consumers lost to businesses no less than 10 times the amount that business lost to criminals in 1971.[54] Clearly, tighter control of corporations, and of business practices, is warranted.

Strong, equitable controls will be necessary at all levels of world commerce (local, state, regional, national and international). As the President's Council on Environmental Quality has pointed out, in the absence of equitable and forceful controls, competitive pressures will make it impossible for any individual under-developed city, state, region, or nation, however self-disciplined, to withdraw from the corporate network which is presently sweeping the world toward immense new spasms of old-style industrialization. The developed nations like our own can individually initiate controls on themselves any time they decide to.

At the international level, we believe that—for all its obvious shortcomings—the United Nations must be the agency through which international controls will develop. We simply haven't time to create alternative agencies. UN leader U Thant said in 1969:

> I do not wish to seem overdramatic, but I can only conclude from the information that is available to me as Secretary General, that the Members of the United Nations have perhaps 10 years left in which to subordinate their ancient quarrels and launch a global partnership to curb the arms race, to improve the human environment, to defuse the population explosion, and to supply the required momentum to development efforts. If such a glo-

bal partnership is not forged within the next decade, then I very much fear that the problems I have mentioned will have reached such staggering proportions that they will be beyond our capacity to control.[55]

But given that we cannot yet control huge corporate forces that appear to be pushing the world blindly toward the final bang or whimper, what should an individual *do*? And what *can* an individual do?

At the most personal level, we can reduce our own environmental damage and the environmental damage that is done to us by our food supply. To avoid the worst problems from expensive processed, embalmed and contaminated supermarket foods, keep your diet as varied as possible and as fresh and unprocessed as possible. Learn about vitamins and proteins and trace elements and healthy food. Learn to get off the top of the food chain. To do this read books by Michael Jacobson, Adelle Davis, Roger J. Williams, Henry Schroeder and Frances Moore Lappe. See the selected bibliography at the back of this book.

As you reform your food habits, you will begin to become a "contrasumer" (Al Fritsch's term). It's an important changeover that doesn't necessarily mean you start living like a hippie, but you become conscious of your place in the environment and how your personal lifestyle and mind-set may be the root of the nation's problems. The public interest research organization we run in Albuquerque, New Mexico, has published what we consider an important little book on reducing your personal pollution. It's called *Our Corner of the Earth, A New Mexican's Guide to Environmental Living*.[56] The author of the study is Lynne Behnfield of New Mexico Citizens for Clean Air & Water. She surveyed middle-class Albuquerque residents and found they use an average of eighty different chemicals in their homes. Ms. Behnfield spent more than a year and compiled a guide which tells you how to complete household cleaning chores with five basic chemicals (such as vinegar, baking soda, etc.), none of which is environmentally damaging in household quantities. Ms. Behnfield has outlined it all in a "chore chart"—how to keep house without spreading destructive chemicals throughout the biosphere. We *can* live sane, sound lives in modest comfort, all of us. Become a contrasumer, or at least read Al Fritsch's book and give it serious thought.

As Anne and Paul Ehrlich have said, probably the most realistic attitude to adopt toward the world situation during the next thirty years is this: Hope for the best but prepare for the worst. If you want to follow up on that philosophy and think about what it means, read the Erhlichs' 1975 book, *The End of Affluence*.[57] It contains good sources of information on how to survive when the going gets rough and how to prepare your family against that time. Throughout the industrial world the going may get rough at any time, unexpectedly and for numerous complex reasons. People should work against but should be prepared for major collapses of significant parts of the industrial system as our explosively growing human systems rapidly encounter the subtle but finite limits of the biosphere.

We'll never solve fundamental problems until the public knows what's going on. Public awareness of environmental problems will require effective new public education programs, especially through the mass media: radio, movies and newspapers in addition to TV. The outlines of our fundamental problems are now known; these must be communicated to the people at every opportunity. Americans have repeatedly demonstrated their ability to solve problems creatively once they understand what the problems are.

The MIT-sponsored Study of Critical Environmental Problems in 1970 said:

> No change of life-style will occur unless the need for change is broadly recognized and accepted. It remains the responsibility of those who study environmental problems and understand their dimensions to convey this need to the public.[58]

Perhaps most important, SCEP recommended that we develop *new means of educating people*, new ways of provoking informed discussion of and action on critical environmental problems. Here we need a factor of 10 expansion of the federal program authorized by Congress in the Environmental Education Act of 1970 (P.L. 91-516). In addition to making environmental education accessible to the people, we must educate all professional people to new awareness of the interrelatedness of earth's ecological life-support systems. All professional people—lawyers, doctors, engineers, architects, teachers, transportation planners, builders, public health

administrators, publishers, clergymen, financiers, and journal-
ists . . . we've left out a lot—*all* professional people—must come
to understand, in broad outline at least, the very real limitations
of earth. The Executive Committee of the Club of Rome wrote in
1972:

> We are convinced that realization of the quantitative re-
> straints of the world environment and of the tragic conse-
> quences of an overshoot is essential to the initiation of
> new forms of thinking that will lead to a fundamental re-
> vision of human behavior and, by implication, of the en-
> tire fabric of present-day society.[59]

In our own personal lives, the way we've chosen is to try to expand a diverse movement for change, called public interest research. The founders of the modern phase of this movement were
Mr. Nader, Dr. Commoner, Dr. Ehrlich and their many colleagues.
Now their ideas are being taken up by people nationwide. Ourselves, we operate a public interest research organization called
Southwest Research and Information Center in Albuquerque, New
Mexico. We also publish a monthly magazine called *The Workbook*,
which is a fully indexed catalog of sources of information about environmental, social and consumer problems. It is aimed at helping
people in small towns and cities across America gain access to vital
information that can help them assert control over their own lives.

The good people of America are in the doldrums right now, but
America has gone through periods of excessive corruption before.
The administration of Ulysses Grant at the close of the Civil War,
the administration of Warren Harding at the close of World War I
and now the Nixon-Ford administration following the close of still
another major period of war, represent our worst.

Still, despite these displays of our society's most criminal and
most nearly Fascist tendencies, across the land small groups of citizens have been and are still working in their various ways to provide a world characterized by economic justice; clean air and water
and uncontaminated soils; abundant, safe, minimally-polluting
supplies of energy; rapid, safe, convenient, efficient and low-cost
public transportation systems; convenient recreational areas; clean,
quiet and safe streets; shops and stores offering a variety of durable, safe, and inexpensive consumer goods; thriving small-town

and regional-scale economies; readily available, adequate child-care facilities; inexpensive and humane health care; participatory democracy; wholesome fairly-priced food; land reform and aid to family farms; decent housing for all; easily accessible education services; widely available legal help; readily accessible scientific and technical tools and advice; low-cost, rapid access to the whole world's cumulative knowledge; science-based land-use planning; justice in the courts; attractive, clean workplaces; and meaningful work for all. We hope that in some small way this book can promote citizen action to serve this vast nameless movement for change, because we agree most emphatically with Barry Commoner, who in 1970 said:

> I'm very upset by people who say there's a crisis of survival and if we don't make it by 1972 we ought to give up. I think that is a disastrous approach. I take the position that we are in charge. We're human beings. We have the resources, we have the knowledge, by God we will do it. . . .
>
> One zoology graduate [student] at a Canadian university got one pickerel and measured the mercury content— it was way above the standards. *Chemical Engineering News* says he torpedoed a half billion dollar a year industry as a result of that one measurement [and] a letter he wrote to the Canadian Ministry; the entire fishing in the area closed down. Two chemical companies have been pinpointed as the source of the mercury, and the Canadian Ministry has now proposed that they pay recompense to the fishermen. *One guy. One guy and a fish.* Now I don't buy this business that we're helpless.[60]

Notes

INTRODUCTION

1. Council on Environmental Quality, *Toxic* . . . , p. 1.
2. Anonymous, "Military. . . ."
3. Hall, pp. 31–35.
4. Harrison Brown, "Human . . . ," pp. 121–22. See also Brown, *The Challenge.* . . .
5. Platt, pp. 1115–21.
6. Commoner, "Soil . . . ," p. 6.
7. *Ibid.*, p. 10, and Commoner, "A Reporter at Large," p. 90.
8. Platt, p. 1115.
9. Study of Critical Environmental Problems, p. 126.

CHAPTER 1

1. Cloud, p. 67. See also Mason, *Principles.* . . .
2. Ehrlich, *Population Resources* . . . , pp. 161–62.
3. Deevey, p. 87.
4. Schroeder, *Pollution, Profits* . . . , p. 69. See also Roger Williams, *Nutrition* . . . , pp. 1–44.
5. See, for example, Frieden, "Biochemical . . . ," pp. 42–46.
6. Hutchinson, p. 10.
7. Deevey, p. 83.
8. Bolin, p. 49.
9. Rabinowitch cited in Asimov, p. 110. See also Ryther, "Is the World's . . . ," pp. 374–75, and Broecker, pp. 1537–38.
10. Woodwell, p. 26.

11. Hutchinson, pp. 3, 5.
12. *Ibid.*, p. 6.
13. Schroeder, "Trace Metals . . . ," p. 281. See also Shaw, p. 210.
14. Hutchinson, p. 11.
15. Commoner, "Soil and . . . ," p. 9.
16. Hutchinson, p. 11.
17. Asimov, p. 157.
18. Guyton, pp. 21–23.
19. Lenihan, pp. 145–46.
20. D'Alonzo, p. 71.
21. Schutte, p. 17.
22. *Ibid.*, p. 13.
23. Asimov, pp. 232–33.
24. Kench cited in Schutte, pp. 33, 201.
25. Schutte, p. 17.
26. Schroeder, "Trace Metals . . . ," pp. 268–70.
27. Schutte, pp. 23–24, 32–33, 111.
28. Cloud, p. 67.
29. Hutchinson, p. 11.
30. Asimov, p. 277.
31. Schroeder, "Editorial," p. 219.
32. George K. Davis, pp. 229–30.
33. Schroeder, "Trace Metals . . . ," p. 220.
34. Schroeder, "Inorganic . . . ," pp. 546–47. See also Schroeder, *The Trace Elements.* . . .
35. Asimov, pp. 220–30.
36. Maugh, pp. 253–54.
37. Underwood, pp. 281ff.
38. Carnes, pp. 29–34. See also Pories, "Trace . . . ," pp. 114–33. See especially O'Dell, pp. 134–40.
39. Schroeder, "Trace Elements . . . ," pp. 973–77. See also Schroeder, "Possible . . . ," pp. 59–67; Schroeder, "Pollution . . . ," pp. 25, 30; Schutte, p. 122; Perry, pp. 179–95; Masironi, pp. 687–97; D'Alonzo, pp. 71–79; Schroeder, "Abnormal . . . Vanadium," pp. 1047–71; Henzel, pp. 83–99; Mertz, pp. 86–95; Baumslag, pp. 23–25; Hueper, pp. 475–81; Borneff, pp. 22–29. See also National Academy; Coon, "Naturally . . . ," pp. 55–59; Coon, "Food . . . ," pp. 103–8.
40. Roger J. Williams, *Nutrition* . . . , p. 45.
41. Jacobson, *Nutrition* . . . , p. 3.
42. Cotzias, p. 4. See also Schroeder, "Trace Metals . . . ," pp. 259–303.
43. Ehrlich, *Population Resources* . . . , p. 346.
44. Schroeder, "Abnormal . . . Chromium," p. 948. See also Frieden, "The Chemical . . . ," p. 52.
45. Asimov, pp. 86–87.

CHAPTER 2

1. Nef, p. 211. See also Mumford, p. 155 and the entire introductory section.
2. Rockefeller 3d, pp. 45–53. See also David Freeman.
3. Singer, p. 108.
4. Smith, p. 7.
5. Hubbert, p. 33.
6. Asimov, pp. 19, 34–60.
7. Ekirch, pp. 176–77.
8. Council on Environmental Quality, *Toxic* . . . , p. iv, emphasis added. See also p. 3.
9. Council on Environmental Quality, *Environmental . . . Second . . .*, p. 224.
10. Mrak, p. 357. See also Crossland, "The Hazards . . . ," pp. 35–38.
11. Anonymous, "Mercury in the Ecosystem," pp. 1ff.
12. Study, p. 22. See also Donella H. Meadows, p. 30.
13. Newill quoted in Lannan, p. A-12. See also Anonymous, cover story, pp. 289–91.
14. Schroeder, *Pollution* . . . ," p. 29.
15. Schroeder, "Metals in the Air," pp. 18–24, 29–32.
16. Crossland, "Drinking . . . ," pp. 11–19.
17. Griporopoulos, p. 252.
18. Anonymous, "Water . . . ," pp. 18–19.
19. Crossland, "Drinking . . . ," pp. 11–19.
20. *Ibid.*
21. Ribicoff, p. 1.
22. Jacobson, *Nutrition* . . . , p. 52.
23. President's, pp. 75ff.
24. *Ibid.*
25. Salant, pp. 72–78.
26. Study, p. 303. See also Institute, p. 48.
27. Hutchinson, p. 11.
28. Ryckman, p. 155.

CHAPTER 3

1. Schroeder, "Essential . . . Zinc . . . ," p. 179.
2. McCaull, "Building . . . ," p. 10. See also Schroeder, *The Trace Elements* . . . , p. 105.
3. Schroeder, *The Trace Elements* . . . , p. 145.

4. Schroeder, "Cadmium," p. 165. See also Committee on Resources, p. 138.
5. Schroeder, "Inorganic . . . ," p. 559.
6. Hambridge, pp. 868–74.
7. Underwood, pp. 267–70.
8. Malin, p. 754.
9. McCaull, "Building . . . ," p. 10.
10. National Kidney.
11. Underwood, pp. 267–70.
12. Schroeder, "Cadmium, Chromium . . . ," p. 574.
13. Schroeder, "Cadmium," p. 165.
14. Council on Environmental Quality, *Environmental . . . Second . . . ,* p. 100.
15. Anonymous, "Toxicity . . . ," p. 348.
16. Evans, pp. 173–74. See also Harrison, p. 302.
17. Dunphy, p. 22.
18. Schlaepfer, pp. 556–64. See also Townshend, pp. 68–69; Heath, pp. 592–93; Friberg, "Proteinuria . . . ," pp. 32–36; Parijek, pp. 111–12; Gabbiani, pp. 154–60.
19. Epstein, pp. 56, 57. See also Anonymous, "Cadmium Toxicity," pp. 1206–7.
20. Paterson, p. 295. Hardy, pp. 8–13.
21. Anonymous, "Cadmium Poisoning," pp. 392–93.
22. W. E. Davis.
23. Council on Environmental Quality, *Toxic Substances,* p. 7.
24. Hyder.
25. MacNamara, p. 1.
26. U.S. Department of Health, Education, and Welfare, July 13, 1971. See also February 8, 1971; February 13, 1970. See FDA news release, March 26, 1971. See also Byrnes, p. 1.
27. Carper, "Nader . . . ,"
28. Schroeder, *Pollution . . . ,*" p. 26.
29. Cristensen, p. 11. See Browning, p. 100. See also Beton, p. 297; Tucker, p. [174].
30. Moser, p. 4.
31. Schroeder, "Cadmium, Chromium . . . ," p. 570–77.
32. Schroeder, "Possible . . . ," p. 65.
33. Schroeder, "Trace Metals . . . ," pp. 284–86, 295.
34. Perry, pp. 179–95.
35. Carroll, "The Relationship . . . ," p. 178.
36. Masironi, pp. 687–97. See also Schroeder, *Pollution . . . ,* pp. 25–26; Morton, p. 537.
37. Schroeder, *Pollution . . . ,* p. 26. See also McCaull, "Building . . . ," p. 6.

38. Schroeder, *Pollution* . . . , pp. 25–26.
39. Schroeder, *The Trace Elements* . . . , p. 107.
40. Carper, "Too Much . . . ," p. A-3. See also U.S. Geological Survey.
41. Carroll, "Trace . . . ," pp. 230–31.
42. Lagerwerff, p. 353.
43. Council on Environmental Quality, *Toxic* . . . , p. 11. See also Schroeder, "Cadmium: Uptake . . . ," p. 819; Schroeder, "Essential . . . Zinc . . . ," p. 195; Lagerwerff, p. 352; Schroeder, "Cadmium, Chromium . . . ," pp. 575–76.
44. Schroeder, "Cadmium as a . . . ," p. 652. See also Institute of Ecology, p. 150.
45. Meranger, p. 364.
46. Bird, "Further . . . ," p. 41.
47. Schroeder, *The Trace Elements* . . . , p. 105.
48. Schroeder, "Essential . . . Zinc . . . ," pp. 195, 206, 208. See also Schroeder, "Abnormal . . . Cadmium," pp. 241, 243–45.
49. Nilsson, p. 49. See also Friberg, *Cadmium.* . . .
50. Tsuchiya, "Causation . . . , Part I," pp. 181–94. See also Tsuchiya, "Causation . . . , Part II," pp. 195–211; and Malin, p. 754.
51. Institute of Ecology, pp. 96, 150–51.
52. McCaull, "Building . . . ," p. 41.

CHAPTER 4

1. Schroeder, *Pollution* . . . , p. 24. See also Council on Environmental Quality, *Toxic* . . . , p. 11.
2. *Ibid.*, p. 23.
3. Ehrlich, *Population Resources* . . . , p. 135.
4. *Ibid.*
5. Patterson, p. 352.
6. Underwood, p. 442. See also Hall, p. 32.
7. Schroeder, "Some Prospects . . . ," p. 23.
8. Lave, p. 723.
9. Roger J. Williams, *Nutrition Against* . . . , p. 11.
10. Lambou, "Hazards . . . ," p. 1.
11. Patterson, pp. 349–50.
12. Institute of Ecology, p. 144.
13. Patterson, pp. 349–50.
14. Gilfillan, p. 56.
15. Patterson, p. 348.
16. Fritsch, "Response. . . ." See also Bird, "Lead . . . ," p. 39; Bird, "New Lead . . . ," p. 42.
17. Stupka, pp. 4–5.

18. Institute of Ecology, p. 146. See also Fristedt, p. 64c.
19. Lewis, p. 186. See also Lane, p. 6.
20. Lagerwerff, pp. 353–54. See also Schroeder, *Pollution* . . . , p. 21; Ullman, p. 361.
21. Anonymous, "Lead in the Air," p. 529.
22. Hall, p. 32.
23. Anonymous, "Lead in the Air," p. 529.
24. Hill, "Report . . . ," p. 28.
25. Council on Environmental Quality, *Environmental . . . Second . . .* , p. 226.
26. *Ibid.*
27. Lambou, "Hazards . . . ," p. 9; see chart.
28. Ulmer, pp. 7–8. See also Anonymous, "Lead in Air," pp. 653–54.
29. Greengard, p. 271.
30. Anonymous, "Lead in Air," pp. 653–54. See also Hall, p. 33.
31. Passow, pp. 203, 208.
32. Lagerwerff, p. 353. See also Stupka, p. 2; Schuck, p. 238; John, p. 1203; Motto, p. 231.
33. Hall, p. 32. See also Ehrlich, *Population Resources* . . . , p. 135.
34. Mintz, p. F-1. See also Anonymous, "EPA Low . . . ," p. 13.
35. Atkins, pp. 26–27.
36. Stupka, p. 2.
37. Lacasse, p. 238. See also Harley, p. 225.
38. Anonymous, "Poisoned Pastures," pp. 1–3. See also Chow, "Lead Isotopes . . . ," pp. 510–11; Rabinowitz.
39. Anonymous, reported in the Washington *Post,* December 16, 1973, p. A-9.
40. United Press International, May 11, 1974, p. B-12.
41. Lagerwerff, p. 357.
42. Kopp, p. 66.
43. Ullman, p. 362.
44. Hall, p. 32. See also Underwood, p. 442.
45. Lee, p. 288–90.
46. Dimmick, pp. 81–82. See also Altshuler, pp. 323–37; Dautrebande, pp. 178–87.
47. Bingham, pp. 83–90.
48. McCaull, "Already Too . . . ," p. 39.
49. Chow, "Lead Aerosols . . . ," p. 740.
50. Schroeder, *Pollution . . . .* " p. 22. See also Schroeder, "Inorganic . . . ," p. 568.
51. Brent, pp. 61–70. See also Wilma Castle.
52. Karstad, pp. 355–60.
53. Morgan, p. 454. See also Voors, p. 97.

54. Stupka, pp. 134–35.
55. Moser, pp. 33–109.
56. Anonymous, "Lead in Air," p. 654.
57 *Ibid.*, p. 653. See also U.S. Department of Health, Education, and Welfare, "Public Health . . . ," p. 44.
58. Institute of Ecology, p. 147.

CHAPTER 5

1. Hartung. See also Friberg, *Mercury in the* . . . ; Wallace, *Mercury in the.* . . .
2. Joselow, pp. 1745–47.
3. Kurland, pp. 370–91. See also McAlpine, September 20, 1958, pp. 629–31; October 11, 1958, p. 802.
4. Clarkson, p. 352. See also Bidstrup, pp. 59–61.
5. Jalili, p. 303.
6. Curley, pp. 65–67.
7. Allen L. Hammond, pp. 788–89.
8. Smart, pp. 2–36; Lofroth, "Birds Give Warning," pp. 10 ff; Novick, "A New . . . ," pp. 2 ff.
9. Westoo, "Methylmercury Compounds in Fish," pp. 11–12. See also Westoo, "Methylmercury Compounds in Animal . . . ," pp. 75–90. See Anonymous, "Cooperative . . . ," pp. 1–3; Burrows, pp. 1127–30; Westoo, "Methylmercury as Percentage . . . ," pp. 567–68; Bache, p. 951.
10. Dean, p. 1044.
11. Jensen, pp. 753–54; Wood, pp. 173–74; Uchida, pp. 181–87. See also Kondo, pp. 137–41.
12. Interagency Committee, p. 9. See also Fagerstrom, pp. 121–22.
13. Lambou, "Proposed Environmental . . . ," p. 64.
14. Spangler, pp. 192–93.
15. George, p. 2.
16. Curley, pp. 65–67. See also Waldron, p. A-1; Anonymous, "Follow-up: Organic . . . ," pp. 4–7.
17. Fimreite, "Mercury . . . ," pp. 119–31; "Extent . . . ," p. 28. See also Klein.
18. U.S. Department of the Interior, "Secretary Hickel . . . ." See also Anonymous, "Mercury . . . ," pp. 24 ff; U.S. Geological Survey, p. 4; Joensuu, pp. 1027–28.
19. Abelson, p. 237.
20. Reed, p. 24.
21. Hill, "Mercury Hazard . . . ," p. 1.

22. Anonymous, "Good-by . . . ," p. 93d. See also U.S. Department of Health, Education, and Welfare, February, 5, 1971; Lyons, "Mercury . . . ," p. 1.
23. Anderson, p. A-7.
24. Lambou, "Proposed . . . ."
25. Grant, p. 15. See also Newberne, pp. 311–35.
26. Clark.
27. Lambou, personal communication.
28. Lambou, "Proposed . . . ," p. 10. See also pp. 12, 30.
29. Nelson, pp. 1–69. See also pp. 20–23.
30. Lambou, "Proposed . . . ," pp. 9, 87.
31. *Ibid.*, p. 87.
32. *Ibid.*, p. 111, emphasis added.
33. *Ibid.*, pp. 10, 88.
34. *Ibid.*, p. 31.
35. Bird, "Canadians . . . ," p. 9.
36. Associated Press, "F.D.A. Limit . . . ," p. 14; Cohn, "Ecologist . . . ," p. A-2. See also Herdman; Associated Press, "Woman on Swordfish . . . ," p. A-1.
37. Wright, p. 771.
38. Newberne, pp. 311–335.
39. Nelson, pp. 9, 12–13, 16.
40. Lyons, "High Levels . . . ," p. 26; Freeman, pp. 7–10.
41. McAlpine, September 20, 1958, p. 631.
42. Clarkson cited in Newberne, p. 331.
43. J.J., pp. 112–13.
44. Grant, "Legacy . . . ," p. 43.
45. Ramel, pp. 208–30. See also pp. 231–54; Malling, pp. 7–9.
46. Skerfving, pp. 133–39.
47. United Press International, "Chromosome . . . ," p. G-8.
48. Spyker, pp. 621–23.
49. George, p. 5.
50. Cohn, "U.S. Study . . . ," p. A-3.
51. Anonymous, "Mercury: What . . . ," pp. 4–6.
52. Anonymous, "The Mercury Menace," p. 47; Cohn, "Senators . . . ," p. A-1; Associated Press, "Woman on . . . ," p. A-1; Goldwater, "Symptoms . . . ," p. 38; Lyons, "Senators . . . ," p. 1. See also Korns, pp. 21–23.
53. Robert C. Spear.
54. Montague, *Mercury.*
55. Lambou, "Proposed . . . ," p. 12, emphasis added. See also Interagency Committee, p. 8.
56. Bird, "Mercury Is Found . . . ," p. 1; Anonymous, "90 Year . . . ," p. A-7.

57. Anonymous, "Japan Fears . . . ," p. a.
58. Anonymous, "Survey on Mercury," p. 3. See also Anonymous, "Japan Toughens . . . ," pp. 6–7.
59. Saha, p. 141.
60. Council on Environmental Quality, *Environmental Quality . . . Second . . .* , pp. 225–26.
61. Anonymous, "Mercury in the Ecosystem," pp. 1 ff.
62. Anonymous, "Extent of Pollution . . . ," p. 28.
63. Anonymous, "Mercury in the Ecosystem," p. 16.
64. Meadows, p. 82.

CHAPTER 6

1. Wallick. See also Davidson; Page; Scott.
2. Council on Environmental Quality, *Environmental Quality . . . Second . . .* , p. 129.
3. Davidson, p. 5. See also Page.
4. Anonymous, *The President's . . .* , p. 59.
5. *Ibid.*, p. 63. See also Gordon, p. 12.
6. U.S. Department of Health, Education, and Welfare, *Facts . . .* , p. 31.
7. Anonymous, *The President's . . .* , p. 111.
8. *Ibid.*, p. 1.
9. *Ibid.*, p. 111.
10. Oil, Chemical and Atomic Workers International Union (Atlanta, Georgia), p. 53.
11. Golz, pp. 281–82.
12. Davidson, p. 154.
13. Oil, Chemical . . . (Houston, Texas), p. 24. See also Oil, Chemical . . . (Tulsa, Oklahoma), p. 48.
14. Anonymous, *The President's . . .* , pp. 3, 87.
15. Oil, Chemical . . . (Kenilworth, New Jersey), pp. 57–59.
16. Hunter. See also Browning; Gagager; Plunkett; Sax; Medical Committee . . . .
17. Lane, p. 6.
18. Anonymous, "Dermatitis," p. 173. Mastromatteo, p. 129.
19. Bech, p. 1.
20. Lane, p. 7.
21. Fristedt, p. 64d.
22. Bech, pp. 242, 251. See also Hamdi, p. 126; Schroeder, "Abnormal . . . Nickel," p. 61.
23. Dinman, pp. 137–41.
24. Nelson, pp. 105–18.

25. Anonymous, "Spectrum," p. 21.
26. Schroeder, *The Trace Elements* . . . , p. 87.
27. Commoner, "Workplace . . . ," pp. 15–20. See also Anonymous, "Ecology Concerns . . . ," p. E-1; Scott, p. 62.
28. Berman, "Diagnosis . . . ," p. 16. See also Hueper, p. 205; Fristedt, pp. 64d-64e; Mastromatteo, p. 130; Buchanan, p. 115.
29. Council on Environmental Quality, *Toxic* . . . , p. 3.
30. Anonymous, *The President's* . . . , p. 101; Glenn Paulson, quoted in Oil, Chemical . . . (Tulsa, Oklahoma), p. 50.
31. Davidson, p. 6.
32. Scott, p. 84.
33. Oil, Chemical . . . (Fort Wayne, Indiana), p. 64.
34. McKiever, p. i.
35. *Ibid.*, p. 23.
36. Scott, p. 41.
37. Chow, "Lead Aerosols . . . ," p. 740.
38. Bingham, p. 85.
39. Tsuchiya, "Lead Exposure . . . ," pp. 181–82, 185–86.
40. Hall, p. 35.
41. Anonymous, "Lead in Air," p. 654.
42. Tsuchiya, "Proteinuria . . . ," p. 880. See also Potts, p. 61.
43. Selikoff, "III. Community . . . ," p. 1661.
44. *Ibid.*, p. 1662.
45. Selikoff, "Asbestos Air . . . ," p. 12.
46. Selikoff, "III. Community . . . ," p. 1658 and fns. 8 and 12.
47. Brodeur, p. 40, quoting Dr. William Stewart, Surgeon General of the United States.
48. Nicholson, quoted in Oil, Chemical . . . (Houston, Texas), p. 8.
49. Selikoff, "Asbestos Air . . . ," p. 2.
50. U.S. Department of Health, Education, and Welfare, *Criteria* . . . , pp. I-1.
51. Oil, Chemical . . . (Houston, Texas), p. 13.
52. *Ibid.*, p. 12.
53. Selikoff, "Asbestos Air . . . ," p. 9.
54. Anonymous, "Beryllium," pp. 35–36.
55. Aldridge, p. 375; Royston, p. 391; Lieben, p. 494. See also Commoner, *Workplace* . . . , pp. 15–20.
56. Griggs, pp. 842–43.
57. Selikoff, "III. Community . . . ," p. 1658.
58. *Ibid.*
59. Selikoff, "Asbestos Air . . . ," p. 2.
60. *Ibid.*
61. *Ibid.*, p. 11.

62. *Ibid.* See also Hueper, p. 204.
63. Montague, "Fiber . . . ," pp. 6–9.
64. Wagner, p. v.
65. Dixon, pp. 141, 143. See also Anonymous, *The President's . . .* , p. 122; Cannon, p. 27; Cralley, p. 453; Speil, pp. 166–208.
66. Oil, Chemical . . . (Houston, Texas), p. 12.
67. Schroeder, "Airborne . . . ," pp. 83–88.
68. Hueper, p. 207.
69. Barry Commoner, "Workplace . . . ," p. 16.
70. Oil, Chemical . . . (Fort Wayne, Indiana), p. 2; Oil, Chemical . . . , (Houston, Texas), p. 15.

CHAPTER 7

1. Wise.
2. Ehrlich, *Population Resources . . .* , p. 118.
3. Srole, p. 332. See also National Clearinghouse. . . .
4. Lombardo, p. 10.
5. *Ibid.*, p. 9.
6. *Ibid.*, pp. 9–12, 15, 18. See also Schroeder, "Relations Between . . . ," p. 590; Tipton, p. 5; Meigs, pp. 363–65; and Morgan, p. 455.
7. Handlin.
8. Anonymous, "King Cotton . . . ," pp. 40–42.
9. Advisory Committee, p. 5. See also Rockefeller 3d, p. 29.
10. Meigs, p. 363. See also Rockefeller 3d, p. 218.
11. Hicks, p. 9.
12. Anonymous, "Reports on Cases . . . ," p. 61. See also Lepow, p. 354; Hall, p. 33; Committee on Environmental . . . , pp. 918–21.
13. Greengard, p. 269.
14. See Chapter 9.
15. Lin-Fu, pp. 1289–93.
16. Meigs, pp. 363–69.
17. Quoted in Hicks, p. 9.
18. Sayre, pp. 167–70.
19. Hall, p. 33.
20. Aaronson and Kohl, "Spectrum," November, 1972, p. 23.
21. *Ibid.* See also Ehrlich, *Population Resources . . .* , p. 134; Liversedge, p. 16.
22. Perlstein, p. 298. See also Thompson, pp. 1, 12.
23. Greengard, p. 275.
24. Lepow, p. 354. See also Chisholm, pp. 1–38; Barltrop, p. 234.
25. Barnes, pp. A-1, A-15.

26. Anonymous, "Reports on Cases . . . ," p. 61.
27. Jacobziner, pp. 279–81.
28. Barnes, p. A-15. See also Magidson, pp. 10–13.
29. Hicks, p. 9.
30. Berman, "Diagnosis . . . ," p. 17.
31. Lepow, p. 355.
32. Kubota, p. 792. See also Sayre; Thompson.
33. Barnes, p. A-15.
34. Magidson, p. 13.
35. Avery Taylor, p. 10; Spear, p. 46. See also Thompson.
36. Barbara Campbell, p. 39.
37. Jacobziner, p. 285. See also Barltrop, p. 233.
38. Anonymous, "Reports on Cases . . . ," p. 61.
39. Avery Taylor, p. 10.
40. Hall, p. 32. See also Berman, "The Biochemistry . . . ," p. 287; A. M. G. Campbell, p. 68; and Perlstein, p. 298.
41. Craig, p. 5.
42. Hall, p. 32.
43. Perlstein, p. 298.
44. Hicks, p. 9.
45. Perlstein, p. 270.
46. Fristedt, p. 64c.
47. Greengard, p. 272.
48. Chisholm, pp. 2, 33. See also Greengard, p. 275.
49. Perlstein, p. 292. See also Berg, pp. 44–53.
50. Craig, p. 5.
51. Bazell, pp. 130–31. See also Lawrence Altman, p. 45; Associated Press, "Animal . . . ," p. E-7; Associated Press, "Detroit Zoo . . . ," p. B-3.
52. Hicks, p. 9.

Chapter 8

1. Oliver Goldsmith.
2. Caldwell.
3. Institute of Ecology, p. 99.
4. *Ibid.*, p. 118.
5. *Ibid.*, pp. 113–17. See also Rasmussen, p. 303; Mrak, p. 45.
6. Mrak, p. 491.
7. Lagerwerff, p. 345.
8. Mrak, pp. 104–5. See also Warren, p. 18; Carson, p. 231.
9. Bowen, p. 166.
10. Miller.

11. Institute of Ecology, pp. 113–14, 117.
12. Mrak, p. 59.
13. Institute of Ecology, pp. 119, 121.
14. Study of Critical, pp. 158–59. See also Institute of Ecology, p. 119.
15. McKeown, p. S8592.
16. Montague, "Ralph Nader . . . ," pp. 58–94.
17. Bowsher, p. 19. See also U.S. Department of Agriculture, *Changes in . . .* , p. 8.
18. Easley, p. vi.
19. McGovern, pp. S1802–8.
20. Bowsher, p. 19.
21. U.S. Senate, *Parity Returns . . .* , pp. 14, 25.
22. McKeown, p. S8585.
23. Durost, p. 3.
24. Institute of Ecology, pp. 25–26.
25. Holdren, p. 285.
26. Institute of Ecology, p. 92.

Chapter 9

1. Thomas R. A. Davis, pp. 43–45.
2. R. J. Williams, "The Concept . . . ," p. 287. See also Roger J. Williams, *Biochemical. . . .*
3. Thomas R. A. Davis, pp. 53–54.
4. *Ibid.*, p. 53.
5. *Ibid.*
6. Roger J. Williams, *Nutrition Against . . .* , p. 39. See also Jacobson, *Eater's . . .* , pp. 6–11.
7. Thomas R. A. Davis, pp. 50–51.
8. Maugh, pp. 253–54.
9. Rasmussen, pp. 122–26. See also Williams, *Nutrition Against . . .* , p. 5.
10. Underwood, p. 9.
11. Roger J. Williams, *Nutrition Against . . .* , pp. 14–15.
12. Anonymous, "Health Care . . . ," p. 1.
13. U.S. Bureau of the Census, p. 71.
14. Roger J. Williams, *Nutrition Against . . .* , pp. 14–16.
15. Asimov, pp. 216–37.
16. Vitamin Information Bureau, "Zinc . . . ," pp. 2–3.
17. Pories, "Health Effects . . . ," pp. 77–78.
18. Underwood, p. 209; Pories, "Trace Elements . . . ," pp. 114–33.; Vitamin Information . . . , p. 11.

19. Pories, pp. 114–33.
20. Schroeder, "Trace Metals . . . ," p. 297.
21. Roger J. Williams, *Nutrition Against* . . . , pp. 67–91. See also Hudson, pp. 30–43; Schroeder, *The Trace Elements* . . . ; Roger J. Williams, *Nutrition in a* . . . , pp. 70, 71, 100.
22. Vitamin Information Bureau, p. 4.
23. Pories, pp. 114–33. See also McCaull, "Building a . . . ," p. 9; Parizek, "Effect . . . ," p. 1036, and Parizek, untitled communication, p. 1037; Underwood, p. 277.
24. Vitamin Information Bureau, p. 3.
25. Hoff, p. 3.
26. Viets, p. 91.
27. U.S. Department of Agriculture, *Changes in* . . . , p. 8.
28. Schutte, p. 177.
29. Murphy, p. 6.
30. Schutte, pp. 66, 148–49, 193.
31. Keeney, pp. 80–81.
32. Council on Environmental Quality, *Environmental . . . Second* . . . , p. 107. See also Wadleigh.
33. Schroeder, "Editorial . . . ," p. 221.
34. Murphy, p. 17.
35. Underwood, p. 241.
36. Hambridge, pp. 868–74. See also Swenerton, pp. 62–64. See also Schroeder, *The Trace Elements* . . . , pp. 36–37, 51.
37. Hambridge, p. 873.
38. Schroeder, "Abnormal . . . Chromium," p. 941.
39. Underwood, pp. xiii, 253–64.
40. Schroeder, "Cadmium, Chromium . . . ," p. 576.
41. U.S. Department of Health, Education, and Welfare, *Diabetes* . . . , pp. 1–2.
42. Underwood, p. 254. See also Schroeder, "Cadmium, Chromium . . . ," p. 572.
43. Schroeder, "Cadmium, Chromium . . . ," p. 577.
44. Schroeder, *Pollution* . . . , pp. 73–74.
45. Schroeder, "Cadmium, Chromium . . . ," p. 578.
46. Schroeder, "Trace Elements in Degenerative . . . ," p. 975. See also Schroeder, *Pollution* . . . , p. 74.
47. Schroeder, *The Trace Elements* . . . , pp. 78–128. See also Roger J. Williams, *Nutrition in a* . . . , p. 58.
48. Schroeder, "Cadmium, Chromium . . . ," p. 577.
49. Atkins, p. 29. See also Schroeder, *Pollution* . . . , p. 77; Henzel, pp. 83–86; and Schroeder, "Trace Elements . . . ." pp. 976–77.
50. Schroeder, *The Trace Elements* . . . , p. 68.

51. Schroeder, "Abnormal . . . Chromium," p. 954. See also Mertz, "Problems . . . ," pp. 166–67; Metz, "The Role . . . ," p. 92; Schroeder, "Some Prospects . . . ," p. 22.
52. Atkins, p. 27; See also Schutte, pp. 123–24; Hudson, pp. 32–43.
53. Underwood, p. 260.
54. Schroeder, "Cadmium, Chromium . . . ," p. 577. See also Doisy, p. 76–77; Mertz, p. 91.
55. Schroeder, *Pollution* . . . , p. 77.
56. Schroeder, "Cadmium, Chromium . . . ," p. 576.
57. Lombardo, p. 7.
58. Schroeder, "Some Prospects . . . ," p. 24. See also Underwood, pp. 261–62; Mertz, p. 88; Schroeder, *Trace Elements* . . . , p. 72.
59. Davis, p. 53.

CHAPTER 10

1. Lave, pp. 723–33.
2. Hodgson, pp. 589–97. See also Anonymous, correspondence, June, 1971, pp. 548–50.
3. Anonymous, "Cancer . . . ," pp. 97–98. See also Anonymous, "Cancer Prevention," pp. 1278–79.
4. United Press International, "Retardation . . . ," p. A–2.
5. Rockefeller 3d, p. 131. See also Scanlon, pp. 135–41; Propping, pp. 49–55.
6. U.S. Department of Health, Education, and Welfare, *Facts* . . . , p. 28.
7. Asimov, pp. 252–53.
8. *Ibid.* See also pp. 173–78.
9. Anonymous, "Longevity . . . ," p. 5.
10. Cohn, "'Alarming' . . . ," p. A–1.
11. Anonymous, "Recent . . . ," pp. 3–6.
12. U.S. Department of Health, Education, and Welfare, *Facts* . . . , p. 24.
13. Lester R. Brown, "Human . . . ," p. 95.
14. Ehrlich, *Population Resources* . . . , p. 322.
15. Study of Critical . . . , p. 61.
16. *Ibid.,* p. 60.
17. Council on Environmental Quality, *Environmental* . . . *Second* . . . , p. 100.
18. Study of Critical . . . , pp. 162–63.
19. Council on Environmental Quality, *Environmental* . . . *Second* . . . , p. 215.

20. Study of Critical . . . , p. 67.
21. Starr, p. 11. See also Schneider, pp. 1–10.
22. Natusch, pp. 202–4. See David Freeman, p. 23.
23. Singer, p. 110. See also Study of Critical . . . , pp. 11–12, 54–55, 98.
24. See Holdren, p. 288.
25. Borchert, pp. 1–22.
26. Associated Press, "Growing Ice . . . ," p. F–8.
27. *Ibid.*
28. Ogburn, p. 10.
29. David Freeman, p. 51.
30. *Ibid.*, p. 56.
31. Ehrlich, *Population* . . . , p. 190. See also Study of Critical . . . , pp. 242, 267; Singer, p. 112; Anonymous, "NAS Probes . . . ," p. 7.
32. Council on Environmental Quality, *Environmental . . . Second . . .* , p. 236; Institute of Ecology, pp. 230–31; Study of Critical . . . , p. 124.
33. Ryther, pp. 72–76. See also Bascom, pp. 16–25.
34. Institute of Ecology, pp. 70, 225; see also p. 260. See also Study of Critical . . . , p. 148.
35. Gasiewicz, pp. 726–28.
36. Council on Environmental Quality, *Environmental . . . Second . . .* , p. 107.
37. *Ibid.*
38. Meranger, pp. 363–64. See also Lambou, "Hazards . . . ," p. 18.
39. Schroeder, "Essential . . . Copper," p. 1025. Arthur, p. 1277; Stupka, pp. 115, 120, 142, 147–48; Lambou, "Hazards . . . ," p. 15; Institute . . . , p. 70.
40. Ehrlich, *Population Resources . . .* , pp. 161–62.
41. Commoner, *Science and . . .* , p. 127.

Chapter 11

1. Study of Critical . . . , p. 126.
2. *Ibid.*, p. 158. See also Council on Environmental Quality, *Environmental Quality . . . Second . . .* , p. 264; Goldsmith, *Blueprint . . . .*
3. Ogburn, pp. 3–36.
4. Bagley, pp. 26–32. See also Landsberg, pp. 333–77.
5. Study of Critical . . . , p. 115.
6. Meadows, p. 51.
7. Study of Critical . . . , p. 118.
8. Ehrlich, "Impact . . . ," p. 1214.
9. Anonymous, "A Gloomy . . . ," p. 58.

10. *Ibid.*
11. Wilson, pp. 2–4.
12. Anonymous, "A Gloomy . . . ," p. 58.
13. Rockefeller 3d, p. 65.
14. Anonymous, "Spectrum," November, 1974, p. 24.
15. Ehrlich, "Impact . . . ," p. 1215.
16. Ryther, "Photosynthesis . . . ," pp. 72–76.
17. Committee on Resources . . . , p. 119.
18. *Ibid.*, pp. 6, 141.
19. *Ibid.*, pp. 125, 153.
20. Meadows, pp. 56–60.
21. Committee on Resources . . . , p. 124.
22. *Ibid.*, p. 130.
23. *Ibid.*, p. 128. See also Ridker, p. 9; Anonymous, *Business* . . . , p. 43; Anonymous, *Chemical & Engineering* . . . , p. 7; Anonymous, "Y.S. Resource . . . ," p. A–4; Anonymous, "Risky . . . ," pp. 64–65; Wade, pp. 185–86; Rockefeller 3d, p. 74.
24. Anonymous, "How Wrong . . . ," pp. 44 ff.
25. Anonymous, "Why Utilities . . . ," pp. 49–51.
26. Anonymous, "How Wrong. . . ," p. 46.
27. Main, p. 116.
28. Boffey, p. 1555.
29. Main, p. 118.
30. Meyers, pp. 123–25 ff. See also Anonymous, "Some 250 Los Alamos . . . ," p. 8; Board of Directors . . . .
31. Starr, p. 10.
32. Summers, p. 103.
33. Novick, "Toward . . . ," pp. 32–40.
34. Main, p. 196.
35. Fitzgerald, p. 227.
36. Novick, "Toward . . . ," pp. 32–40.
37. Hoffman, pp. 45–52. See also Sternglass; Gofman; Curtis; Metzger; Novick, *The Careless*. . . .
38. Bowen, p. 198. See also pp. 103, 115, 172.
39. McPhee, p. 94. See also Parts II and III.
40. *Ibid.*, p. 120.
41. *Ibid.*, pp. 102–5.
42. *Ibid.*, p. 125.
43. McPhee, Part III, p. 94.
44. *Ibid.*, p. 93.
45. Belter, p. 6.
46. Hambleton, p. 4.
47. Ehrlich, *Population Resources* . . . , p. 137.

48. Lanouette, p. 4. See also Anonymous, "A New Star . . . ," p. 6.
49. Personal communication, summer, 1973.
50. Weinberg, p. 36.
51. Novick, *The Careless* . . . , p. 86. See also Lapp, "Thoughts . . . ," p. 12E; Lapp, "Safety," p. 21.
52. Associated Press, "Scientists . . . ," p. 53.
53. Ripley, pp. 1, 54. See also 61. Forbes, pp. 40–47; Ford, pp. 3–9 ff.
54. O'Toole, "Environmentalists . . . ," p. A–2.
55. Fitzgerald, pp. 236–39.
56. Lapp, "Safety," p. 21; emphasis added.
57. Weinberg, p. 33; Lovins, pp. 29–36.
58. Carpenter, p. 1320.
59. Ehrlich, *Population Resources* . . . , pp. 296–98.
60. Committee on Resources . . . , p. 117.
61. *Ibid.*, pp. 120–21. See also Ehrlich, *Population Resources* . . . , p. 61.
62. Anonymous, *U.S. News & World* . . . , pp. 21, 23.
63. Committee on Resources . . . , p. 117.
64. O'Toole, "Taiwan Getting . . . ," p. A–9.
65. Gilinsky, pp. 11–17. See also Anonymous, "And Now For . . . ," pp. 14–15.
66. Weinberg, p. 36.
67. Meadows, pp. 193–94.

CHAPTER 12

1. Ahmed, pp. 6–12. See also Eigner, pp. 15–18.
2. Holdren, p. 290.
3. Anonymous, "Costs of Pollution . . . ," p. 10.
4. Council on Environmental Quality, *Toxic* . . . .
5. Kneese, *Pollution, Prices.* . . .
6. Christensen.
7. Uetz, pp. 31–39.
8. Holdren, pp. 282–292.
9. Van Dyne. See also Odum; Garvey; Kneese, *Economics.* . . . ; Kneese, *Environmental.* . . .
10. Theodore Taylor.
11. Council on Environmental Quality, *Toxic* . . . . See also Castleman; Montague, "Fiber . . . ," pp. 6–9.
12. Harrison Brown, "Human . . . ," p. 122.
13. Meadows, p. 108.
14. Singer, p. 114.

15. Ehrlich, *Population Resources* . . . , pp. 69–75.
16. Meadows, pp. 67–68.
17. Thurow, p. 3.
18. *Ibid.,* p. 4.
19. Anonymous, "Costs of Pollution. . . ."
20. Kneese, "Environmental . . . ," p. 7.
21. Rockefeller 3d , pp. 113–14.
22. Theodore Taylor.
23. Altman, pp 53–64.
24. Detwyler.
25. Schneider, pp. 1–10. See also Haefele.
26. NSF/NASA Solar . . . .
27. Anonymous, "Solar . . . ," pp. 41–42.
28. Rosenblatt, pp. 99–111.
29. Loferski, pp. 3–10.
30. Anonymous, "Solar . . . ."
31. David Freeman, pp. 42–43.
32. Cox. See also Dieges, pp. 20–21; Aaronson, "Black Box . . . ," pp. 10–18.
33. William Hammond, p. 88.
34. Dials, pp. 18–24, 30–37.
35. David Freeman; Seamons.
36. Lester Brown, *World.* . . .
37. Olson.
38. Odum.
39. Anonymous, "Technical Aspects. . . ."
40. Holdren.
41. Reilly.
42. Kemeny.
43. Nova, pp. 1065 ff.
44. Brent, p. 65.
45. L.H.N., p. 377.
46. *Ibid.,* p. 378.
47. Zielinski, p. 1.
48. Wellford, pp. 191–92, 108.
49. U.S. Senate, *Chemicals* . . . , pp. 76–103.
50. Feuer, p. 394.
51. Peirce, pp. 1–15.
52. Nader.
53. Sax.
54. Biderman, p. 49, quoting Senator Philip Hart.
55. Meadows, p. 17.
56. Behnfield.

57. Ehrlich, *The End.* . . .
58. Study of Critical . . . , pp. 159–60.
59. Meadows, p. 190.
60. Mills, p. 18.

Bibliography

AARONSON, TERRI. "Black Box." *Environment,* Vol. XIII (December, 1971), pp. 10–18.

———, and GEORGE KOHL. "Spectrum." *Environment,* Vol XIV (November, 1972), p. 23.

ABELSON, PHILIP H. "Methyl Mercury." *Science,* Vol. CLXIX (July 17, 1970), p. 237.

ADVISORY COMMITTEE ON INTERGOVERNMENTAL RELATIONS. *A Commission Report, Urban and Rural America: Policies for Future Growth.* Washington, D.C.: U.S. Government Printing Office, 1968.

AHMED, A. KARIM. "Unshielding the Sun—by Diminishing Ozone—Human Effects." *Environment,* Vol. XVII (April/May, 1975), pp. 6–12.

ALDRIDGE, W. N., J. M. BARNES, and F. A. DENZ, "Experimental Beryllium Poisoning." *British Journal of Industrial Pathology,* Vol. XIX (1949), pp. 375–389.

ALTMAN, LAWRENCE K. "Zoo Cat's Death Starts Lead-Poison Study." New York *Times,* June 15, 1971, p. 45.

ALTMAN, M., MARIA TELKES, and M. WOLF. "The Energy Resources and Electric Power Situation in the United States." *Energy Conversion,* Vol. XII (May 31, 1971), pp. 53–64.

ALTSCHULER, BERNARD, EDWARD D. PALMAS, and NORTON NELSON. "Regional Aerosol Deposition in the Human Respiratory Tract." In Charles Norman Davies, *Inhaled Particles and Vapours.* Vol. II. New York: Pergamon Press, 1967.

ANAS, R. E. "Mercury in Fur Seals." In Donald R. Buhler, ed. *Mercury in the Western Environment.* Corvallis, Oregon: Continuing Education Publications [Ext. Hall Annex, Corvallis, Oregon 97331], 1973.

ANDERSON, JACK, "Conservationists Beg EPA to Free Mercury Report." Albuquerque *Journal,* April 16, 1972, p. A-7.

ANONYMOUS, "And Now for a Little Diversion. . . ." *Environment,* Vol. XIV (October, 1972), pp. 14–15.

————. "Beryllium." *Environment,* Vol. XVI (April, 1974), pp. 35–36.

————. "Cadmium Poisoning." *British Journal of Medicine,* May 13, 1967, pp. 392–93.

————. "Cadmium Toxicology." *The Lancet,* June 14, 1969, pp. 1206–7.

————. "Cancer and Ecology." *Newsweek,* July 10, 1972, pp. 97–98.

————. "Cancer Prevention." *The Lancet,* June 27, 1970, pp. 1378–79.

————. "Cooperative Mercury Pollution Research." *SFI* [Sport Fishing Institute] *Bulletin* , No. 257 (August, 1974), pp. 1–3.

————. "Costs of Pollution Abatement." *New Mexico Business,* Vol. XXV (May, 1972), pp. 7–10, 25–26.

————. "Dermatitis." *Industrial Medicine and Surgery,* Vol. XXVI (April, 1957), p. 173.

————. "Ecology Concerns Factory Workers." Denver *Post,* March 19, 1972, p. E-1.

————. "EPA Low Lead Rules Called Wasteful." *Chemical & Engineering News,* Vol. LI (December 10, 1973), p. 13.

————. "EPA Spells Out Drinking Water Standards." *Chemical & Engineering News,* Vol. LII (March 24, 1975), p. 6.

————. "Escalating Prices of Nonferrous Metals." *Business Week,* March 10, 1973, p. 43.

————. "ES&T Interview—Vaun Newill." *Environmental Science and Technology,* Vol. VI (July, 1972), pp. 289–91.

————. "Extent of Pollution Much Wider Than Once Seen." *Medical Tribune,* June 9, 1971, p. 28.

————. "Follow-up: Organic Mercury Poisoning—Alamogordo, New Mexico." *Center for Disease Control Neurotropic Diseases Surveillance* [No. 1, Mercury Poisoning]. Atlanta, Georgia: U.S. Department of Health, Education, and Welfare, and Public Health Service, Health Services and Mental Health Administration, Center for Disease Control, March 15, 1971.

————. "A Gloomy Forecast for Grain-Hungry Countries." *Business Week,* September 14, 1974, p. 58.

————. "Good-by to Swordfish." *Newsweek,* May 17, 1971, p. 93d.

————. "Health Care Priorities Backward, Says Researcher." *National Health Federation Bulletin,* Vol. XVIII (April, 1972), p. 1.

————. "How Wrong Forecasts Hurt the Utilities." *Business Week,* February 13, 1971, pp. 44 ff.

————. "Japan Fears Spread of Mercury Disease," New York *Times,* July 7, 1974, p. 2.

———. "Japan Toughens Pollution Control Stance." *Chemical & Engineering News*, April 2, 1973, pp. 6-7.

———. "King Cotton Blasts Off." In Jack Hayes, ed. *Contours of Change* [Yearbook of Agriculture, 1970; House Document 91–254, Ninety-first Congress, Second Session]. Washington, D.C.: U.S. Government Printing Office, 1970.

———. "Lead in Air." *British Medical Journal*, September 18, 1971, pp. 653–54.

———. "Lead in the Air." *Environmental Science and Technology*, Vol. III (June, 1969), p. 529.

———. "Longevity at 65—International Trends." *Statistical Bulletin* [of the Metropolitan Life Insurance Company], Vol. LI (July, 1970), p. 5.

———. "Mercury in the Air." *Environment*, Vol. XIII (May, 1971), pp. 24 ff.

———. "Mercury in the Ecosystem." *CBNS Notes* [Center for the Biology of Natural Systems, Washington University, St. Louis], Vol. IV (September–October, 1971), pp. 1–4, 16.

———. "The Mercury Menace." *Newsweek*, May 31, 1971, p. 47.

———. "Mercury, Watch That Tuna." *Medical World News*, May 18, 1973, pp. 4–6.

———. "Metal Group Assesses Impact of Shortages." *Chemical & Engineering News*, Vol. LII (March 4, 1974), p. 7.

———. "Military Units Here Work on New Bomb." Albuquerque *Journal*, October 1, 1970, p. E-6.

———. "NAS Probes Fate, Effects of Ocean Oil." *Chemical & Engineering News*, Vol. LII (May 20, 1974), p. 7.

———. "A New Star Has Appeared in the Antinuclear Firmament." *Nucleonics Week*, Vol. XV (February 28, 1974), p. 6.

———. "90-Year-Old Tuna Samples Show Mercury." Albuquerque *Journal*, August 22, 1971, p. A-7.

———. "Poisoned Pastures." *California's Health*, Vol. XXVIII (July/ August, 1970), pp. 1–3.

———. *The President's Report on Occupational Safety and Health*. Washington, D.C.: U.S. Government Printing Office, May, 1972.

———. "Recent Mortality Trends for Emphysema and Other Chronic Respiratory Diseases." *Statistical Bulletin* [of the Metropolitan Life Insurance Company], Vol. LI (August, 1970), pp. 3–6.

———. "Reports on Cases of Lead Poisoning Rise 300% Here." New York *Times*, February 23, 1971, p. 61.

———. "Risky Race for Minerals." *Time* , January 28, 1974, pp. 64–65.

———. "Solar Cell Offers 21% Efficiency With Gas." *Electronics*, Vol. XLVII (May 29, 1975), pp. 41–42.

————. "Some 250 Los Alamos Scientists, Among Others, Constructively Question Nuclear Power." *Nucleonics Week*, March 13, 1975, p. 8.

————. "Spectrum." *Environment*, Vol. XVI (November, 1974), pp. 21, 24.

————. "Survey on Mercury." *Tokyo Municipal News*, Vol. XXIII (September, 1973), p. 3.

————. "Toxicity of Cadmium." *The Lancet*, August 15, 1964, pp. 348–49.

————. "U.S. Resource Crisis Seen." Washington *Post*, April 22, 1973, p. A-4.

————. "Water Contaminated Throughout U.S." *Chemical & Engineering News*, Vol. LII (April 28, 1975), pp. 18–19.

————. "Why U.S. Risks War for Indochina—It's the Key to Control of All Asia." *U.S. News & World Report*, April 16, 1954, pp. 21, 23.

————. "Why Utilities Can't Meet Demand." *Business Week*, November 29, 1969, pp. 49–51.

ARTHUR, JOHN W., and EDWARD N. LEONARD. "Effects of Copper on *Gammarus pseudolimnaeus, Physa integra,* and *Campeloma decisum* in Soft Water." *Journal Fisheries Research Board of Canada*, Vol. XXVII (1970), pp. 1277–83.

ASIMOV, ISAAC. *The Intelligent Man's Guide to the Biological Sciences.* New York: Washington Square Press, 1968.

ASSOCIATED PRESS. "Animal Lead Buildup May Indicate Danger to Humans." Albuquerque *Journal*, June 18, 1971, p. E-7.

————. "Detroit Zoo Probing Film Pack Toxic Level." Albuquerque *Journal*, July 24, 1971, p. B-3.

————. "F.D.A. Limit on Mercury Called High." New York *Times*, December 29, 1970, p. 14.

————. "Growing Ice Cap May Be Trend." Albuquerque *Journal*, May 15, 1974, p. F-8.

————. "Scientists Warn of a 'Disaster' from Nuclear Reactor System." New York *Times*, July 27, 1971, p. 53.

————. "Woman on Swordfish Diet Gets Mercury Poisoning." Albuquerque *Journal*, May 21, 1971, p. A-1.

ATKINS, HARRY. "Elemental Traces." *The Sciences*, Vol. XI (May, 1971), pp. 26–27.

BACHE, C. A., W. H. GUTEMANN, "Residues of Total Mercury and Methylmercuric Salts in Lake Trout as a Function of Age." *Science*, Vol. CLXXII (May 28, 1971), pp. 951–52.

BAGLEY, GEORGE R. "The Need for Environmental Management." In *Proceedings, Second Annual State-Wide Land-Use Planning Symposium.* Albuquerque, N.M., 1970.

BAKIR, F., S. F. DAMLUJI, L. AMIN-ZAKI, *et al.* "Methylmercury Poisoning in Iraq." *Science*, Vol. CLXXXI (July 20, 1973), pp. 230–41.

BARLTROP, DONALD. "Lead Poisoning." *Archives of Disease in Childhood*, Vol. XLVI, No. 46 (1971), pp. 233–35.

BARNES, ANDREW. "Lead Poison: Threat to Poor." Washington *Post*, February 28, 1971, pp. A-1, A-15.

BASCOM, WILLARD. "The Disposal of Waste in the Ocean." *Scientific American*, Vol. CCXXXI (August, 1974), pp. 16–25.

BAUMSLAG, NAOMI, and PAUL KEEN. "Trace Elements in Soil and Plants and Antral Cancer." *Archives of Environmental Health*, Vol. XXV (July, 1972), pp. 23–25.

BAZELL, ROBERT J. "Lead Poisoning: Zoo Animals May Be First Victims." *Science*, Vol. CLXXIII (July 9, 1971), pp. 130–31.

BECH, A. O., M. D. KIPLING, and J. C. HEATHER. "Hard Metal Disease." *British Journal of Industrial Medicine*, Vol. XIX (1961), pp. 239–52.

BEHNFIELD, LYNNE K. *Our Corner of the Earth: A New Mexican's Guide for Environmental Living.* Albuquerque, N.M.: Southwest Research and Information Center [P.O. Box 4524, Albuquerque, N.M. 87106], 1975.

BELTER, WALTER G. "Ground Disposal of Radioactive Waste in an Expanding Nuclear Power Industry." Typescript. Washington, D.C.: U.S. Atomic Energy Commission [paper presented to the annual meeting of the American Institute of Mining, Metallurgical and Petroleum Engineers, Washington, D.C., February 18, 1969].

BERG, J. M., and M. ZAPPELLA. "Lead Poisoning in Childhood with Particular Reference to Pica and Mental Sequelae." *Journal of Mental Deficiency Research*, Vol. I (1964), pp. 44-53.

BERMAN, ELEANOR. "The Biochemistry of Lead: Review of the Body Distribution and Methods of Lead Determination," *Clinical Pediatrics*, Vol. V (May, 1966), pp. 287–91.

————. "Diagnosis of Poisoning." *Journal of Occupational Medicine*, Vol. XIII (January, 1971), pp. 14–18.

BETON, D. C., G. S. ANDREWS, H. J. DAVIES, *et al.* "Acute Cadmium Fume Poisoning." *British Journal of Industrial Medicine*, Vol. XXIII (1966), pp. 292–301.

BIDERMAN, PAUL L. "Consumer Class Actions Under the New Mexico Fair Practices Act." *New Mexico Law Review*, Vol. IV (November, 1973), p. 49.

BIDSTRUP, P. LESLEY. "Clinical Symptoms of Mercury Poisoning in Man." *Biochemical Journal*, Vol. CXXX (1972), pp. 59–61.

BINGHAM, EULA. "Trace Amounts of Lead in the Lung." In Delbert D. Hemphill, ed. *Trace Substances in Environmental Health—III.* Columbia: University of Missouri Press, 1970.

BIRD, DAVID. "Canadians Wary on Mercury Rate." New York *Times*, February 17, 1971, p. 9.

————. "Further Oyster-Cadmium Study Urged." New York *Times*, April 12, 1971, p. 41.

————. "Lead in City Air Causing Concern." New York *Times*, March 4, 1971, p. 39.

_____. "Mercury Is Found in Fish Dating Back to '27." New York *Times*, January 6, 1971, p. 1.

_____. "New Lead-Poisoning Source Sought." New York *Times*, January 20, 1971, p. 42.

Board of Directors of New Mexico Citizens for Clean Air and Water. *A Nuclear Power Policy*. Santa Fe: New Mexico Citizens for Clean Air and Water [100 Circle Drive, Santa Fe, N.M. 87501], March 18, 1975.

BOFFEY, PHILIP M. "Energy Crisis: Environmental Issue Exacerbates Power Supply Problem." *Science*, Vol. CLXVIII (June 26, 1970), pp. 1554–59.

BOLIN, BERT. "The Carbon Cycle." In Dennis Flanagan, ed. *The Biosphere*. San Francisco: W. H. Freeman, 1970.

BORCHERT, JOHN R. "The Dust Bowl in the 1970s." *Annals of the Association of American Geographers*, Vol. LXI (March, 1971), pp. 1–22.

BORNEFF, J., F. SELENKA, H. KUNTE, and A. MAXIMOS. "Experimental Studies on the Formation of Polycyclic Aromatic Hydrocarbons in Plants. *Environmental Research*, Vol. II (1968), pp. 22–29.

BOWEN, H. J. M. *Trace Elements in Biochemistry*, New York: Academic Press, 1966.

BOWSHER, PRENTICE. "CPR [Center for Political Research] Department Study, the Agriculture Department." *National Journal*, Vol. II (January 3, 1970), p. 19.

BOYLE, THOMAS J. "Hope for the Technological Solution." *Nature*, Vol. CCLXV (September 21, 1973), pp. 127–28.

BRENT, ROBERT L. "Protecting the Public from Teratogenic and Mutagenic Hazards." *Journal of Clinical Pharmacology* Vol. XII(February–March, 1972), pp. 61–70.

BRODEUR, PAUL. *Asbestos and Enzymes*. New York: Ballantine, 1972.

BROECKER, WALLACE S. "Man's Oxygen Reserves." *Science*, Vol. CLXVIII (June 26, 1970), pp. 1537–38.

BROWN, HARRISON. *The Challenge of Man's Future*. New York: Viking Press, 1954.

_____. "Human Materials Production as a Process in the Biosphere." In Dennis Flanagan, ed. *The Biosphere*. San Francisco: W. H. Freeman, 1970.

BROWN, LESTER R. "Human Food Production as a Process in the Biosphere." In Dennis Flanagan, ed. *The Biosphere*. San Francisco: W. H. Freeman, 1970.

_____. *World Without Borders*. New York, 1972 [Vintage Books ed., 1973].

BROWNING, ETHEL. *Toxicity of Industrial Metals*. 2d ed. New York: Appleton-Century-Crofts, 1969.

BUCHANAN, W. D. *Toxicity of Arsenic Compounds*. Amsterdam: Elsevier, 1962.

BUHLER, DONALD R., ed. *Mercury in the Western Environment*. Corvallis, Oregon: Continuing Education Publications [Ext. Hall Annex, Corvallis, Oregon 97331], 1973.

BURCHELL, ROBERT W., and DAVID LISTOKIN. *The Environmental Impact Handbook*. New Brunswick, N.J.: Rutgers University, Center for Urban Policy Research, 1975.

BURROWS, W. DICKINSON, and PETER KRENKEL. "Studies of Uptake and Loss of Methylmercury 203 by Bluegills (Lepomis macrochirus Raf)." *Environmental Science and Technology*, Vol. VII (December, 1973), pp. 1127–30.

BYRNES, MICHAEL A. "Public Warning." *HEW News*, April 16, 1971, p. 1.

CALDWELL, William A., ed. *How to Save Urban America*. New York: New American Library, 1973.

CAMPBELL, A. M. G., G. HERDAN, W. F. T. TATLOW, and E. G. WHITTLE. "Lead in Relation to Disseminated Sclerosis." *Brain*, Vol. LXXIII (1950), pp. 52–71.

CAMPBELL, BARBARA. "Abrams Asks $3-Million More for a Lead Detection Program." New York *Times*, January 15, 1971, p. 39.

CANNON, HELEN L. "Trace Element Excesses and Deficiencies in Some Geochemical Provinces of the United States." In Delbert D. Hemphill, ed. *Trace Substances in Environmental Health—III*. Columbia: University of Missouri Press, 1970.

CARNES, WILLIAM H., WALTER F. COULSON, DON W. SMITH, and NORMAN WEISSMAN. "The Role of Copper in the Circulatory System." In Delbert D. Hemphill, ed. *Trace Substances in Environmental Health—II*. Columbia: University of Missouri Press, 1969.

CARPENTER, RICHARD A. "Information for Decisions in Environmental Policy." *Science*, Vol. CLXVIII (June 12, 1970), p. 1320.

CARPER, ELSIE. "Nader Unit Cites Cadmium Hazards." Washington *Post*, February 13, 1971.

———. "Too Much Cadmium, Arsenic Found in Water of 12 Areas." Washington *Post*, March 31, 1971, p. A-3.

CARROLL, ROBERT E. "The Relationship of Cadmium in the Air to Cardiovascular Death Rates." *Journal of the American Medical Association*, Vol. CXCVIII (October 17, 1966), pp. 177–79.

———. "Trace Element Pollution of Air." In Delbert D. Hemphill, ed. *Trace Substances in Environmental Health—III*. Columbia: University of Missouri Press, 1970.

CARSON, RACHEL. *Silent Spring*. New York, 1962 [ed. cited is Fawcett paperback].

CASTLE, WILMA. Personal communication. [Dr. Castle was at the University of New Mexico Medical School, Albuquerque, in 1973.]

CASTLEMAN, BARRY I., and ALBERT J. FRITSCH. *Asbestos and You*. Washington, D.C.: Center for Science in the Public Interest, 1974.

CELLARIUS, RICHARD. "Are Solar Power Systems Really Pollution Free?" *Alternative Sources of Energy*, No. 11 (July, 1973), pp. 8–11.

CHISHOLM, J. JULIAN, Jr. "The Use of Chelating Agents in the Treatment of Acute and Chronic Lead Intoxication in Childhood." *Journal of Pediatrics*, Vol. LXXIII (July, 1968), pp. 1–38.

CHOW, TSAIHWA J., and JOHN L. EARL. "Lead Aerosols in Marine Atmosphere." *Environmental Science and Technology*, Vol. II (August, 1969), pp. 737–40.

——. "Lead Isotopes in North American Coal." *Science*, Vol. CLXXVI (May 5, 1972), pp. 510–11.

CHRISTENSEN, F. C., and E. C. OLSON. "Cadmium Poisoning." *Archives of Industrial Health*, Vol. XVI (1957), pp. 8–13.

CHRISTENSEN, HERBERT E., ed. *The Toxic Substances List, 1974 Edition* [HEW Publication No. (NIOSH) 74-134]. Washington, D.C.: U.S. Government Printing Office, 1974.

CLARK, FRANKLIN D. "FDA's Program of Mercury in Foods." In Donald R. Buhler, ed. *Mercury in the Western Environment*. Corvallis, Oregon: Continuing Education Publications [Ext. Hall Annex, Corvallis, Oregon 97331], 1973.

CLARK, WILSON. *Energy for Survival. The Alternative to Extinction*. Garden City, N.Y.: Anchor Press/Doubleday, 1975.

CLARKSON, THOMAS W. "Biochemical Aspects of Mercury Poisoning." *Journal of Occupational Medicine*, Vol. X (July, 1968), 351–55.

CLOUD, PRESTON, and AHARON GIBOR. "The Oxygen Cycle." In Dennis Flanagan, ed. *The Biosphere*. San Francisco: W. H. Freeman, 1970.

COHN, VICTOR. "'Alarming' Increase Is Reported in Cancer Deaths Among Blacks." Washington *Post*, May 18, 1972, p. A-1.

——. "Ecologist Criticizes FDA for 'Dismal' Rules on Mercury." Washington *Post*, December 29, 1970, p. A-2.

——. "Senators Told Woman on Fish Diet Suffered Mercury Damage to Brain." Washington *Post*, May 21, 1971, p. A-1.

——. "U.S. Study Team Urges Curb on Commercial Mercury Use." Washington *Post*, January 15, 1971, p. A-3.

COMMITTEE ON ENVIRONMENTAL HAZARDS, AMERICAN ACADEMY OF PEDIATRICS. "Lead Content of Paint Applied to Surfaces Accessible to Young Children." *Pediatrics*, Vol. XXXXIX (June, 1972), pp. 918–21.

COMMITTEE ON RESOURCES AND MAN [OF THE DIVISION OF EARTH SCIENCES, NATIONAL ACADEMY OF SCIENCES–NATIONAL RESEARCH COUNCIL]. *Resources and Man: Study and Recommendations*. San Francisco: W. H. Freeman, 1969.

COMMONER, BARRY. *The Closing Circle. Man, Nature and Technology*. New York: Knopf, 1971.

——. "A Reporter at Large: The Closing Circle—II." *New Yorker*, October 2, 1971, pp. 44–91.

————. *Science and Survival.* New York: Viking Press, 1966.

————. "Soil and Fresh Water: Damaged Global Fabric." *Environment,* Vol. XII (April, 1970), pp. 4–11.

————. "Workplace Burden." *Environment,* Vol. XV (July/August, 1973), pp. 15–20.

COON, J. M. "Food Toxicology: Safety of Food Additives." *Modern Medicine,* Vol. XXXVIII (November 30, 1970), pp. 103–8.

————. "Naturally Occurring Toxicants in Food." *Food Technology,* Vol. XXIII (1969), pp. 55–59.

COTZIAS, GEORGE C. *Trace Metals: Essential or Detrimental to Life* [Brookhaven Lecture Series BNL 828 (T-323), Biology and Medicine—TID 4500, 24th ed.], No. 26 (April 10, 1963).

COUNCIL ON ENVIRONMENTAL QUALITY. *Environmental Quality: The Second Annual* [1971] *Report of the Council on Environmental Quality.* Washington, D.C.: U.S. Government Printing Office, 1971.

————. *Toxic Substances.* Washington, D.C.: U.S. Government Printing Office, April, 1971.

COX, KENNETH E., ed. *Hydrogen Energy.* Albuquerque: Technology Applications Center, University of New Mexico, 1975.

CRAIG, PAUL P., and EDWARD BERLIN. "The Air of Poverty." *Environment,* Vol. XIII (June, 1971), pp. 2–9.

CRALLEY, L. J., R. G. KENNAN, and J. R. LYNCH. "Exposure to Metals in the Manufacture of Asbestos Textile Products." *American Industrial Hygiene Association Journal,* Vol. XXVIII (September–October, 1967), pp. 452–61.

CROSSLAND, JANICE, and VIRGINIA BRODINE. "Drinking Water." *Environment,* Vol. XI (April, 1973), pp. 11–19.

CROSSLAND, JANICE, and KEVIN P. SHEA. "The Hazards of Impurities." *Environment,* Vol. XV (June, 1973), pp. 35–38.

CURLEY, AUGUST, VINCENT A. SEDLAK, EDWARD F. GIRLING, *et al.* "Organic Mercury Identified as the Cause of Poisoning in Humans and Hogs." *Science,* Vol. 172 (April 2, 1971), pp. 65–67.

CURTIS, RICHARD and ELIZABETH HOGAN. *Perils of the Peaceful Atom: The Myth of Safe Nuclear Power Plants.* New York: Ballantine, 1969.

DAETZ, DOUGLAS, and RICHARD H. PANTELL. *Environmental Modeling: Analysis and Management.* Stroudsburg, Pa.: Dowden, Hutchinson & Ross, 1974.

D'ALONZO, C. ANTHONY, SIDNEY PELL, and ALLAN J. FLEMING. "The Role and Potential Role of Trace Metals in Disease." *Journal of Occupational Medicine,* Vol. V (February, 1963), pp. 71–79.

DAUTREBAND, L., H. BECKMANN, and W. WALKENHORST. "Lung Deposition of Fine Particles." *AMA Archives of Industrial Health,* Vol. XVI (September, 1957), pp. 179–87.

DAVIDSON, Ray. *Peril on the Job.* Washington, D.C.: Public Affairs Press, 1970.

DAVIS, GEORGE K. "Toxicity of the Essential Minerals." In National Academy of Sciences—National Research Council. *Toxicants Naturally Occurring in Foods* [Publication No. 1354]. Washington, D.C.: National Academy of Sciences–National Research Council, 1966.

DAVIS, THOMAS R. A., STANLEY N. GERSHOFF, and DEAN F. GAMBLE. "Review of Studies of Vitamin and Mineral Nutrition in the United States, 1950–1968." *Journal of Nutrition Education,* Vol. I (Fall, 1969 [Supp. I]), pp. 43–45.

DAVIS, W. E., and associates. *National Inventory of Sources and Emissions, Cadmium, Nickel and Asbestos.* Vol. I, *Cadmium* [Publication No. PB 192 250]. Springfield, Va.: U.S. Department of Commerce, National Technical Information Service, 1970.

DEAN, ROBERT B., ROBERT T. WILLIAMS, and ROBERT H. WISE. "Disposal of Mercury Wastes from Water Laboratories." *Environmental Science and Technology,* Vol. V (October, 1971), pp. 1044–45.

DEEVEY, EDWARD S., Jr. "Mineral Cycles." In Dennis Flanagan, ed. *The Biosphere.* San Francisco: W. H. Freeman, 1970.

DETWYLER, THOMAS R., and MELVIN G. MARCUS, eds. *Urbanization and Environment.* Belmont, Calif.: Wadworth Publishing Co., 1972.

DIALS, GEORGE E., and ELIZABETH C. MOORE. "The Cost of Coal." *Environment,* Vol. XVI (September, 1974), pp. 18–24, 30–37.

DIEGES, PAUL. "The Hydrogen-Oxygen Car!" *Alternative Sources of Energy,* No. 11 (July, 1973), pp. 20–21.

DIMMICK, ROBERT L., and ANNE B. AKERS. *An Introduction to Aerobiology.* New York: Wiley-Interscience, 1969.

DINMAN, B. D., "Arsenic: Chronic Human Intoxication." *Journal of Occupational Medicine,* Vol. II (March, 1960), pp. 137–41.

DIXON, J. R., D. B. LOWE, D. E. RICHARDS, and H. E. STOKINGER. "The Role of Trace Metals in Chemical Carcinogenesis—Asbestos Cancers." In Delbert D. Hemphill, ed. *Trace Substances in Environmental Health—II.* Columbia: University of Missouri Press, 1969.

DOISY, R. J., D. H. P. STREETEN, R. A. LEVINE, and R. B. CHODOS. "Effects and Metabolism of Chromium in Normals, Elderly Subjects, and Diabetics." In Delbert D. Hemphill, ed. *Trace Substances in Environmental Health—II.* Columbia: University of Missouri Press, 1969.

DUNPHY, BARRY. "Acute Occupational Cadmium Poisoning: A Critical Review of the Literature." *Journal of Occupational Medicine,* Vol. IX (January, 1967), pp. 22–26.

DUROST, DONALD D., and WARREN E. BAILEY. "What's Happened to Farming." In Jack Hayes, ed. *Contours of Change* [Yearbook of Agriculture, 1970; Ninety-first Congress, Second Session, House Document No. 91-254]. Washington, D.C.: U.S. Government Printing Office, 1970.

EASLEY, L. T. *Food Costs—Farm Prices: A Compilation of Information Relating to Agriculture by the Committee on Agriculture, House of Representatives, Ninety-first Congress, Second Session.* Washington, D.C.: U.S. Government Printing Office, 1970.

EHRLICH, ANNE H., and PAUL R. EHRLICH. *The End of Affluence.* New York: Ballantine, 1975.

EHRLICH, PAUL R., and ANNE H. EHRLICH. *Population Resources Environment.* San Francisco: W. H. Freeman, 1970; 2d ed., 1972.

EHRLICH, PAUL R., and JOHN P. HOLDREN. "Impact of Population Growth." *Science,* Vol. CLXXI (March 26, 1971), pp. 1212–17.

EIGNER, JOSEPH. "Unshielding the Sun—by Diminishing Ozone—Environmental Effects." *Environment,* Vol. XVII (April/May, 1975), pp. 15–18.

EKIRCH, ARTHUR A., Jr. *Man and Nature in America.* New York: Columbia University Press, 1963.

EPSTEIN, SAMUEL S. "Statement of Samuel S. Epstein, M.D., on Adverse Human Effects Due to Chemical Pollutants." In United States Senate, *Chemicals and the Future of Man: Hearings Before the Subcommittee on Executive Reorganization and Government Research of the Committee on Government Operations, United States Senate, Ninety-second Congress, First Session, April 6 and 7, 1971.* Washington, D.C.: U.S. Government Printing Office, 1971.

EVANS, DARREL M. "Cadmium Poisoning." *British Medical Journal,* January 16, 1960, pp. 173–74.

FAGERSTROM, TORBJORN, and ARNE JERNELOV. "Formation of Methyl Mercury from Pure Mercuric Sulphide in Aerobic Organic Sediment." *Water Research,* Vol. V (1971), pp. 121–22.

FEUER, LEWIS S. *The Scientific Intellectual.* New York: Basic Books, 1963.

FIMREITE, NORVALD. "Extent of Mercury Pollution Much Wider Than Once Seen." *Medical Tribune,* June 9, 1971, p. 28.

———. "Mercury Uses in Canada and Their Possible Hazards as Sources of Mercury Contamination." *Environmental Pollution,* Vol. I (1970), pp. 119–31.

FITZGERALD, JOSEPH J., and N. IRVING SAX, eds. *Dangerous Properties of Industrial Materials.* 3d ed. New York: Van Nostrand Reinhold, 1968.

FORBES, IAN A., DANIEL F. FORD, HENRY W. KENDALL, and JAMES T. MACKENZIE. "Cooling Water." *Environment,* Vol. XIV (January–February, 1972), pp. 40–47.

FORD, DANIEL F., and HENRY W. KENDALL. "Nuclear Safety." *Environment,* Vol. XIV (September, 1972), pp. 3–9 ff.

FREEMAN, DAVID, et al. eds. *Exploring Energy Choices: A Preliminary Report of the Ford Foundation's Energy Policy Project.* Washington, D.C.: Energy Policy Project [P.O. Box 23212, Washington, D.C. 20004], 1974.

FREEMAN, DONALD K., ELIZABETH PRICE, WILLIAM PECK, and M. McLAIN. "Investigation of Potential Human Disease Due to Mercury Exposure,

Pribilof Islands, Alaska." *Center for Disease Control Neurotropic Diseases Surveillance No. 1, Mercury Poisoning.* Atlanta, Ga.: U.S. Department of Health, Education, and Welfare, Public Health Service, Health Services and Mental Health Administration, Center for Disease Control, March 15, 1971.

FRIBERG, LARS. "Proteinuria and Kidney Injury Among Workmen Exposed to Cadmium and Nickel Dust." *Journal of Industrial Hygiene and Toxicology,* Vol. XXX (January, 1948), pp. 32–36.

————, MAGNUS PISCATOR, and GUNNAR NORDBERG. *Cadmium in the Environment: A Toxicological and Epidemiological Appraisal* [Publication No. PB 199 795]. Springfield, Va.: U.S. Department of Commerce, National Technical Information Service, April, 1971.

FRIBERG, LARS, and JAROSLAV VOSTAL, eds. *Mercury in the Environment: An Epidemiological and Toxicological Appraisal.* Cleveland Chemical Rubber Co. [18901 Cranwood Parkway, Cleveland, Ohio 44128], 1972.

FRIEDEN, EARL. "Biochemical Evolution of Iron and Copper Proteins, Substances Vital to Life." *Chemical & Engineering News,* Vol. LII (March 25, 1974), pp. 42–46.

————. "The Chemical Elements of Life." *Scientific American,* Vol. CCXXVII (July, 1972), p. 52.

FRISTEDT, B. "Metal Toxicity." In Swedish Natural Science Research Council, eds. *Metals and Ecology* [Ecological Research Committee, Bulletin No. 5, Proceedings of a Symposium Held in Stockholm, March 24, 1969]. Stockholm: Swedish Natural Science Research Council, 1970.

FRITSCH, ALBERT J. *The Contrasumers: A Citizen's Guide to Resource Conservation.* New York: Praeger, 1974.

————, and MICHAEL PRIVAL. "Response to the Environmental Protection Agency's Notice for Additional Health Effects Information Concerning the Use of Leaded Gasoline" (in the *Federal Register* June 14, 1972). Typescript. Washington, D.C.: Center for Science in the Public Interest [1779 Church Street NW, Washington, D.C. 20036], 1972.

GABBIANI, G., D. BAIC, and C. DEZIEL. "Toxicity of Cadmium for the Central Nervous System," *Experimental Neurology,* Vol. XVIII (1971), pp. 154–60.

GAGAGER, W. M., ed. *Occupational Diseases: A Guide to Their Recognition* [Public Health Service Publication No. 1097]. Washington, D.C.: U.S. Government Printing Office, 1964.

GALBRAITH, JOHN KENNETH. *The New Industrial State.* Boston: Houghton Mifflin, 1967.

GARVEY, GERALD. *Energy, Ecology, Economy.* New York: Norton, 1972.

GASIEWICZ, THOMAS A., and FRANK J. DINAN. "Concentration of Mercury in the Manufacture of Fish Protein Concentrate by Isopropyl Alcohol Extraction of Sheepshead and Carp." *Environmental Science and Technology,* Vol. VI (August, 1972), pp. 726–27.

GEORGE, JOHN A., NORVALD FIMREITE, and WILLIAM N. HOLSWORTH. "The Mercury Problem." Unpublished paper delivered by John A. George, April 12, 1970, at Sarnia, Canada, before the Sarnia and District Labor Council.

GILFILLAN, S. C. "Lead Poisoning and the Fall of Rome." *Journal of Occupational Medicine*, Vol. VII (February, 1965), pp. 53–60.

GILINSKY, VICTOR. "Bombs and Electricity." *Environment*, Vol. XIV (September, 1972), pp. 11–17.

GOFMAN, JOHN W., and ARTHUR R. TAMPLIN. *Poisoned Power: The Case Against Nuclear Power Plants*. Emmaus, Pa.: Rodale Press, 1971.

GOLDSMITH, EDWARD, *et al. Blueprint for Survival*. New York: New American Library, 1974.

GOLDSMITH, OLIVER. *The Deserted Village*. London, 1770.

GOLDWATER, LEONARD J. "Mercury in the Environment." *Scientific American*, Vol. CCXXIV (May, 1971), pp. 15–21.

———. "Symptoms of Mercury Poisoning: A Letter to the Editor." *New York Times*, June 2, 1971, p. 38.

GOLZ, HAROLD H., B. DWIGHT CARVER, HARRIET L. HARDY, *et al.* [members of the Committee on Industrial Hygiene and Clinical Toxicology of the Industrial Medical Association]. "Report of an Investigation of Threshold Limit Values and Their Usage." *Journal of Occupational Medicine*, Vol. VIII (May, 1966), pp. 281–82.

GORDON, JEROME, *et al. Industrial Safety Statistics: Re-examination; A Critical Report Prepared for the U.S. Department of Labor*. New York: Praeger, 1971.

GRANT, NEVILLE. "Legacy of the Mad Hatter." *Environment*, Vol. XI (May, 1969), pp. 18–23 ff.

———. "Mercury in Man." *Environment*, Vol. XIII (May, 1971), pp. 2–15.

GREENGARD, JOSEPH. "Lead Poisoning in Childhood: Signs, Symptoms, Current Therapy, Clinical Expressions." *Clinical Pediatrics*, Vol. V. (May, 1966), pp. 269–76.

GRIGGS, KYLE. "Toxic Metal Fumes from Mantle-Type Camp Lanterns." *Science*, Vol. CLXXXI (August 31, 1973), pp. 842–43.

GRIGOROPOULOS, S. G., J. W. SMITH, and J. R. MATTHEWS. "Trace Organic Substances in Missouri Waters." In Delbert D. Hemphill, ed. *Trace Substances in Environmental Health—III*. Columbia: University of Missouri Press, 1970.

GUYTON, ARTHUR C. *Function of the Human Body*. Philadelphia: W. B. Saunders, 1959.

HAEFLE, W., ed. *Proceedings of IIASA Planning Conference on Energy Systems*. Laxenburg, Austria: International Institute for Applied Systems Analysis, 1973.

HALL, STEPHEN K. "Lead Pollution and Poisoning." *Environmental Science and Technology*, Vol. VI (January, 1972), pp. 31–35.

HAMBLETON, WILLIAM W. "Storage of High-Level Radioactive Waste." Typescript. Lawrence: Kansas Geological Survey, University of Kansas, 1971 [a paper presented to the American Association for the Advancement of Science symposium, The Energy Crisis: Some Implications and Perspectives, Philadelphia, December 28, 1971].

HAMBRIDGE, K. MICHAEL, CAROLYN HAMBRIDGE, MARGARET JACOBS, and J. DAVID BAUM. "Low Levels of Zinc in Hair, Anorexia, Poor Growth and Hypogeusia in Children." *Pediatric Research*, Vol. VI (1972), pp. 868–74.

HAMDI, EBTISSAM A. "Chronic Exposure to Zinc of Furnace Operators in a Brass Foundry." *British Journal of Industrial Medicine*, Vol. XXVI (1969), pp. 126–34.

HAMMOND, ALLEN L. "Mercury in the Environment: Natural and Human Factors." *Science*, Vol. 171 (February 26, 1971), pp. 788–89.

―――, WILLIAM D. METZ, and THOMAS H. MAUGH, *Energy and the Future*, Washington, D.C.: American Association for the Advancement of Science, 1973.

HANDLIN, OSCAR. *The Uprooted*. Boston: Little, Brown, 1951; 2d ed., 1973.

HARDY, HARRIET L., and JOHN B. SKINNER. "The Possibility of Chronic Cadmium Poisoning." *Archives of Industrial Health*, Vol. XVI (July, 1957), pp. 8–13.

HARLEY, JOHN H. "Discussion: Sources of Lead in Perennial Ryegrass and Radishes." *Environmental Science and Technology*, Vol. IV (March, 1970), p. 225.

HARRISON, HAROLD E., HENRY BUNTING, NELSON K. ORDWAY, and WILHELM S. ALBRINK. "The Effects and Treatment of Inhalation of Cadmium Aerosols in the Dog." *Journal of Industrial Hygiene and Toxicology*, Vol. XXIX (September, 1947), pp. 302–14.

HARTUNG, ROLF, and BERTRAM D. DINMAN, eds. *Environmental Mercury Contamination*. Ann Arbor: Ann Arbor Science Publishers [P.O. Box 1425, Ann Arbor, Mich. 48106], 1972.

HEATH, J. C., M. R. DINGLE, and M. WEBB. "Cadmium as a Carcinogen." *Nature*, Vol. 193 (February 10, 1962), pp. 592–93.

HENZEL, J. H., B. HOLTMANN, F. W. KEITZER, M. S. DEWEESE, and E. LICHTI. "Trace Elements in Atherosclerosis, Efficacy of Zinc Medication as a Therapeutic Modality." In DELBERT D. HEMPHILL, ed. *Trace Substances in Environmental Health—II*. Columbia: University of Missouri Press, 1969.

HERDMAN, ROGER C. Testimony before Senator Philip Hart's subcommittee of the U.S. Senate Commerce Committee on May 20, 1971.

HICKS, NANCY. "Drive to Stop Lead Poisoning Begins." New York *Times*, October 10, 1970, p. 9.

HILL, GLADWIN. "Mercury Hazard Found Nationwide." New York *Times*, September 11, 1970, p. 1.

―――. "Report on Lead in the Atmosphere of Los Angeles Held Up a Year." New York *Times*, May 13, 1971, p. 28.

HODGSON, T. A. "Short-Term Effects of Air Pollution on Mortality in New York City." *Environmental Science and Technology*, Vol. IV (1970), pp. 589–97.

————, R. E. ECKARDT, *et al.* Correspondence in *Environmental Science & Technology*, Vol. V. (June, 1971), pp. 548–50.

HOFF, GORDON B., CLARK D. LEEDY, and M. DOUGLAS BRYANT. *Fertilizer Guide for New Mexico* [Circular 403]. Las Cruces: New Mexico State University, Cooperative Extension Service, May, 1970.

HOFFMAN, ALLAN R., and DAVID R. INGLIS. "Radiation and Infants." *Bulletin of the Atomic Scientists*, Vol. XXVIII (December, 1972), pp. 45–52.

HOLDREN, JOHN P., and PAUL R. EHRLICH. "Human Population and the Global Environment." *American Scientist*, Vol. LXII (May/June, 1974), pp. 282–92.

HUBBERT, M. KING. "The Energy Resources of the Earth." In Board of Editors of *Scientific American*, eds. *Energy and Power*. San Francisco: W. H. Freeman, 1971.

HUDSON, T. G. FAULKNER. *Vanadium, Toxicology and Biological Significance.* New York: Elsevier, 1964.

HUEPER, W. C. "Occupational Cancer Hazards in American Industries." *Archives of Industrial Hygiene and Occupational Medicine*, Vol. V (March, 1952), pp. 204–8.

————, and W. W. PAYNE. "Carcinogenic Effects of Adsorbates of Raw and Finished Water Supplies." *American Journal of Clinical Pathology*, Vol. XL (May, 1963), pp. 475–81.

HUNTER, DONALD. *The Diseases of Occupations.* 3d ed., Boston: Little, Brown, 1969.

HUTCHINSON, G. EVELYN. "The Biosphere." In Dennis Flanagan, ed. *The Biosphere.* San Francisco: W. H. Freeman, 1970.

HYDER, CHARLES. Personal communication, spring, 1973. [Dr. Hyder is with the Physics Department, University of New Mexico, Albuquerque.]

Institute of Ecology. *Man in the Living Environment.* Madison: University of Wisconsin Press, 1972.

Interagency Committee on Environmental Mercury [State of California]. "Mercury in the California Environment." Mimeographed. Berkeley: California State Department of Public Health, Environmental Health and Consumer Protection Program, 1971.

JACOBSON, MICHAEL F. *Eater's Digest: The Consumer's Factbook of Food Additives.* New York: Doubleday, 1972.

————. *Nutrition Scoreboard: Your Guide to Better Eating.* Washington, D.C.: Center for Science in the Public Interest, 1973.

JACOBZINER, HAROLD. "Lead Poisoning in Childhood: Epidemiology, Manifestations, and Prevention." *Clinical Pediatrics*, Vol. V (May, 1966), pp. 277–86.

JALILI, M. A., and A. H. ABBASI. "Poisoning by Ethyl Mercury Toluene

Sulphonanilide." *British Journal of Industrial Medicine,* Vol. XVIII (1961), p. 303–8.

JENSEN, S., and A. JERNELOV. "Biological Methylation of Mercury in Aquatic Organisms." *Nature,* Vol. 223 (August 16, 1969), pp. 753–54.

J.J. "How Much Metal Is There in Our Waters?" *Environmental Science and Technology,* Vol. VIII (February, 1974), pp. 112–13.

JOENSUU, OIVA I. "Fossil Fuels as a Source of Mercury Pollution." *Science,* Vol. 172 (June 4, 1971), pp. 1027–28.

JOHN, MATT K. "Lead Contamination of Some Agricultural Soils in Western Canada." *Environmental Science and Technology,* Vol. V (December, 1971), pp. 1199–1203.

JOSELOW, MORRIS M. "Editorials—Environmental Negligence: The Mercury Problem." *American Journal of Public Health,* Vol. LXI (September, 1971), pp. 1745–47.

KAHN, HERMAN, and ANTHONY J. WIENER. *The Year 2000.* New York: Macmillan, 1967.

KARSTAD, LARS. "Angiopathy and Cardiopathy in Wild Waterfowl from Ingestion of Lead Shot." *Connecticut Medicine,* Vol. XXXV (June, 1971), pp. 355–60.

KEENEY, D. R., and L. W. JACOBS. "The Remaining Seventy Elements." Typescript. Madison: Department of Soil Science, University of Wisconsin, April, 1971. [This paper will be published in G. Chesters and J. M. Bremner, eds. *Soil Chemistry.* New York: Dekker.]

KEMENY, JOHN G. *Man and the Computer.* New York: Scribner's, 1972.

KENCH, J. E. "Interactions of Metal Ions in Living Cells." In Karl H. Schutte, ed. *Some Aspects of Trace Elements in Nature.* Capetown, South Africa: University of Capetown, 1961.

KLEIN, DAVID H. "Sources and Present Status of the Mercury Problem." In Donald R. Buhler, ed. *Mercury in the Western Environment.* Corvallis, Oregon: Continuing Education Publications [Ext. Hall Annex, Corvallis, Oregon 97331], 1973.

KNEESE, ALLEN V. "Environmental Pollution and the Economy." *Resources,* No. 46 (June, 1974), p. 7.

———, ROBERT U. AYRES, and RALPH C. d'ARGE. *Economics and the Environment: A Materials Balance Approach.* Baltimore: Johns Hopkins Press, 1970.

KNEESE, ALLEN V., and BLAIR T. BOWERS, eds. *Environmental Quality Analysis: Theory and Method in the Social Sciences.* Baltimore: Johns Hopkins Press, 1972.

KNEESE, ALLEN V., and CHARLES L. SCHULTZE, *Pollution, Prices and Public Policy.* Washington, D.C.: Brookings Institution, 1975.

KONDO, T. "Studies on the Origin of the Causative Agent of Minamata Disease." *Journal of the Pharmaceutical Society of Japan,* Vol. LXXXIV (1964), pp. 137–41.

KOPP, JOHN H. "The Occurrence of Trace Elements in Water." In Delbert D. Hemphill, ed. *Trace Substances in Environmental Health—III.* Columbia: University of Missouri Press, 1970.

KORNS, ROBERT F. "The Frustration of Bettye Russow." *Nutrition Today,* Vol. VII (November/December, 1972), pp. 21–23.

KUBOTA, JOE, V. A. LAZAR, and FRED LOSEE. "Copper, Zinc, Cadmium and Lead in Human Blood from 19 Locations in the United States." *Archives of Environmental Health,* Vol. XVI (June, 1968), pp. 788–93.

KURLAND, LEONARD T., STANLEY N. FARO, and HOWARD SIEDLER. "Minamata Disease." *World Neurology,* Vol. I (1960), pp. 370–91.

LACASSE, NORMAN L. "Discussion: Lead in Soils and Plants; Its Relationship to Traffic Volume and Proximity to Highways." *Environmental Science and Technology,* Vol. IV (March, 1970), p. 238.

LAGERWERFF, J. V. "Heavy-Metal Contamination of Soils." In American Association for the Advancement of Science, eds. *Agriculture and the Quality of Our Environment.* Washington, D.C.: American Association for the Advancement of Science, 1967.

LAMBOU, VICTOR. Personal communication.

———, and BENJAMIN LIM. "Hazards of Lead in the Environment, with Particular Reference to the Aquatic Environment." Typescript. Washington, D.C.: U.S. Department of the Interior, Federal Water Quality Administration, August, 1970.

———, *et al.* "Proposed Environmental Protection Agency Position Document Mercury," Mimeographed. Washington, D.C.: Environmental Protection Agency, December, 1971.

LANDSBERG, HANS H., LEONARD L. FISCHMAN, and JOSEPH L. FISHER. *Resources in America's Future.* Baltimore: Johns Hopkins Press, 1963.

LANE, R. E. "Health Hazards from Metals in Industry." *Annals of Occupational Hygiene,* Vol. VIII (1965), pp. 5–11.

LANNAN, JOHN. "Cadmium Becoming Pollution Target." Washington *Evening Star,* August 9, 1970, p. A–12.

LANOUETTE, WILLIAM J. "The Nation's Atomic Garbage Dump?" *National Observer,* February 3, 1973, p. 4.

LAPP, RALPH E. "Safety." *New Republic,* January 23, 1971, p. 21.

———. "Thoughts on Nuclear Plumbing." New York *Times,* December 12, 1971, p. 12–E.

LAVE, LESTER B., and EUGENE P. SESKIN. "Air Pollution and Human Health." *Science,* Vol. 169 (August 21, 1970), pp. 723–33.

LEE, ROBERT E., JR., RONALD K. PATTERSON, and JACK WAGMAN. "Particle-Size Distribution of Metal Components in Urban Air." *Environmental Science and Technology,* Vol. II (1968), pp. 288–90.

LENIHAN, J. M. A. "Technology and Humanity." In Delbert D. Hemphill, ed. *Proceedings University of Missouri's 1st Annual Conference on Trace*

Substances in Environmental Health. Columbia: University of Missouri Press, 1968.

LEPOW, MARTHA. "Discussion of J. Julian Chisholm Jr.'s article entitled Adverse Effects of Inorganic Lead Salts as Related to Various Levels of Absorption of Lead." *Connecticut Medicine,* Vol. XXXV (June, 1971), pp. 353–55.

LEWIS, CHARLES E. "The Toxicology of the Organometallic Compounds—Part I." *Journal of Occupational Medicine,* Vol. II (April, 1960), pp. 183–87.

L.H.N. "Mutagenicity of Chemicals and Drugs." *Connecticut Medicine,* Vol. XXXV (June, 1971), pp. 377–78.

LIEBEN, JAN, and FRANZ METZNER. "Epidemiological Findings Associated with Beryllium Extraction." *American Industrial Hygiene Association Journal,* Vol. XX (December, 1959), pp. 494–98.

LIN-FU, JANE S. "Vulnerability of Children to Lead Exposure and Toxicity." *New England Journal of Medicine,* Vol. CCLXXXIX (December 13, 1973), pp. 1289–93.

LIVERSEDGE, L. A. "Metals and the Central Nervous System." *Annals of Occupational Hygiene,* Vol. VIII (1965), pp. 13–16.

LOFERSKI, JOSEPH J. "Direct Photovoltaic Solar Energy Conversion." In U.S. House of Representatives, *Solar Photovoltaic Energy, Hearings Before the Subcommittee on Energy of the Committee on Science and Astronautics, U.S. House of Representatives, Ninety-third Congress, Second Session, June 5 and 11, 1974* [No. 43]. Washington, D.C.: U.S. Government Printing Office, 1974.

LOFROTH, GORAN. *Methylmercury: A Review of Health Hazards and Side Effects Associated with the Emission of Mercury Compounds into Natural Systems* [Ecological Research Committee Bulletin, No. 4]. Stockholm: Swedish National Research Council, September, 1970.

———, and MARGARET E. DUFFY. "Birds Give Warning." *Environment,* Vol. XI (May, 1969), pp. 10–17.

LOMBARDO, LOUIS V., *et al. Our Urban Environment and Our Most Endangered People* [A Report to the Administrator of the Environmental Protection Agency by the Task Force on Environmental Problems of the Inner City]. Washington, D.C.: Environmental Protection Agency, September, 1971.

LOVINS, AMORY B. "The Case Against the Fast Breeder." *Bulletin of the Atomic Scientists,* March, 1973, pp. 29–36.

LYONS RICHARD D., "High Levels of Mercury Found in Eaters of Seal Meat." New York *Times,* November 6, 1970, p. 26.

———. "Mercury Found High in 89% of Swordfish Tested." New York *Times,* December 24, 1970, p. 1.

———. "Senators Hear Mercury Is Peril to Fish Industry." New York *Times,* May 21, 1971, p. 1.

MacNamara, E. E. "Leachate from Landfilling." *TGC Bulletin* [Technical Guidance Center for Environmental Quality, Department of Environmental Sciences, University of Massachusetts, Amherst], Vol. IV (February, 1972), p. 1.

Magidson, Daniel T. "Half Step Forward." *Environment,* Vol. XIII (June, 1971), pp. 10–13.

Main, Jeremy. "A Peak Load of Trouble for the Utilities." *Fortune,* November, 1969, pp. 116–19, 194–205.

Malin, H. Martin, Jr. "Metals Focus Shifts to Cadmium." *Environmental Science and Technology,* Vol. V (September, 1971), pp. 754–55.

Malling, H. V., J. S. Wassom, and S. S. Epstein. "Mercury in Our Environment." *Newsletter of the Environmental Mutagen Society,* Vol. I, No. 3 (June, 1970), pp. 7–9.

Marx, Leo. "American Institutions and Ecological Ideals." *Science,* Vol. CLXX (November 27, 1970), pp. 945–52.

Masironi, Roberto. "Cardiovascular Mortality in Relation to Radioactivity and Hardness of Local Water Supplies in the USA." *Bulletin of the World Health Organization,* Vol. XLIII (1970), pp. 687–97.

Mason, Brian. *Principles of Geochemistry.* 3d ed. New York: John Wiley & Sons, 1966.

Mastromatteo, Ernest. "Nickel: A Review of Its Occupational Health Aspects." *Journal of Occupational Medicine,* Vol. IX (March, 1967), pp. 127–36.

Maugh, Thomas H., II. "Trace Elements: A Growing Appreciation of Their Effects on Man." *Science,* Vol. CLXXI (July 20, 1973), pp. 253–54.

McAlpine, Douglas. "Minamata Disease." *The Lancet,* October 11, 1958, p. 802.

———, and Shukuro Araki. "Minamata Disease." *The Lancet,* September 20, 1958, pp. 629–31.

McCaull, Julian. "Already Too Late." *Environment,* Vol. XIII (September, 1971), pp. 38–39.

———. "Building a Shorter Life." *Environment,* Vol. XIII (September, 1971), pp. 2–15.

———. "Spectrum," *Environment,* Vol. XV (July/August, 1973), p. 21.

McGovern, George. "Disappearing Farms Cause New American Social Problems." *Congressional Record,* February 1, 1968, pp. S1802–8.

McKeown, Mrs. Gordon. "The Structural Revolution in Agriculture: A Challenge to Production Economics Policymaking; Recommendations for Effective Reform." *Congressional Record,* July 25, 1969, p. S8592.

McKiever, Margaret F., ed. *National Health Survey Findings of Occupational Health Interest* [Public Health Service Publication No. 1418]. Washington, D.C.: U.S. Government Printing Office, 1966.

McPhee, John. "The Curve of Bending Energy [Part I]," *New Yorker,* December 3, 1973, pp. 54–145; [Part II], *New Yorker,* December 10, 1973, pp. 50–108; [Part III], *New Yorker,* December 17, 1973, pp. 60–97.

MEADOWS, DONELLA H., DENNIS L. MEADOWS, JØRGEN RANDERS, and WILLIAM W. BEHRENS III. *The Limits to Growth.* New York: Universe Books, 1972.

MEDICAL COMMITTEE FOR HUMAN RIGHTS. *Health Hazards in the Workplace.* Chicago: Medical Committee for Human Rights [710 South Marshfield, Chicago, Ill. 60612], January, 1972.

MEIGS, J. WISTER, and ELAINE WHITMIRE. "Epidemiology of Lead Poisoning in New Haven Children—Operational Factors." *Connecticut Medicine,* Vol. XXXV (June, 1971), pp. 363–65.

MERANGER, J. C., and E. SOMERS. "Determination of the Heavy Metal Content of Sea-Foods by Atomic-Absorption Spectrophotometry." *Bulletin of Environmental Contamination and Toxicology,* Vol. III (1968), pp. 363–64.

MERTZ, WALTER. "The Role of Chromium in Glucose Metabolism." In Delbert D. Hemphill, ed. *Proceedings University of Missouri's 1st Annual Conference on Trace Substances in Environmental Health.* Columbia: University of Missouri Press, 1968.

METZGER, H. PETER. *The Atomic Establishment.* New York: Simon and Schuster, 1972.

MEYERS, HAROLD B. "The Great Nuclear Fizzle at Old B. & W.," *Fortune,* November, 1969, pp. 123–25 ff.

MILLER, NORTON W., and GEORGE G. BERG, eds. *Chemical Fallout: Current Research on Persistent Pesticides.* Springfield, Ill.: Chas. C. Thomas, 1969.

MILLS, STEPHANIE. "Earth Times Interview: Barry Commoner." *Earth Times,* Vol. I (June, 1970), p. 18.

MINTZ, MORTON. "Lead Content: A Court Issue." Washington *Post,* February 17, 1974, p. F-1.

MONTAGUE, KATHERINE, and PETER MONTAGUE. "Fiber Glass." *Environment,* Vol. XVI (September, 1974), pp. 6–9.

———. "Lead-Poisoned Waterfowl." *Audubon,* Vol. LXXV (May, 1973), pp. 112 ff.

———. *Mercury.* New York: Sierra Club, 1971.

MONTAGUE, PETER. "Ralph Nader and Barry Commoner: Strategies for Public Interest Research, Including Three Original Case Studies." Doctoral dissertation. Albuquerque: University of New Mexico, December, 1971.

MORGAN, JEAN M. "The Consequences of Chronic Lead Exposure." *The Alabama Journal of Medical Sciences,* Vol. V (October, 1968), pp. 454–57.

MORTON, WILLIAM E. "Hypertension and Drinking Water, A Pilot Statewide Ecological Study in Colorado." *Journal of Chronic Diseases,* Vol. XXIII (1971), pp. 537–45.

MOSER, MARVIN, and ARTHUR G. GOLDMAN. *Hypertensive Vascular Disease.* Philadelphia: Lippincott, 1967.

MOTTO, HARRY L., ROBERT H. DAINES, DANIEL M. CHILKO, and CARLOTTA

K. Motto. "Lead in Soils and Plants: Its Relationship to Traffic Volume and Proximity to Highways." *Environmental Science and Technology,* Vol. IV (March, 1970), pp. 231–38.

Mrak, Emil M., *et al. Report of the* [HEW] *Secretary's Commission on Pesticides and Their Relationship to Environmental Health, Parts I and II.* Washington, D.C.: U.S. Government Printing Office, December, 1969.

Mumford, Lewis. *Technics and Civilization.* New York: Harcourt Brace, 1934.

Murphy, L. S., and L. M. Walsh, "Micronutrient Handbook." *Agrichemical Age,* November, 1971, p. 6.

Nader, Ralph, and Mark J. Green, eds. *Corporate Power in America.* New York: Grossman, 1973.

National Academy of Sciences–National Research Council. *Toxicants Occurring Naturally in Foods* [Publication No. 1354]. Washington, D.C.: National Academy of Sciences–National Research Council, 1966.

National Clearinghouse for Mental Health Information, National Institute of Mental Health. *Pollution: Its Impact on Mental Health; a Literature Survey and Review of Research* [U.S. Department of Health, Education, and Welfare Publication No. (HSM) 72–9135]. Washington, D.C.: U.S. Government Printing Office, 1972.

National Kidney Foundation. *Your Kidneys: Master Chemists of the Body.* New York: National Kidney Foundation, n.d.

Natusch, D. F. S., J. R. Wallace, and C. A. Evans Jr. "Toxic Trace Elements: Preferential Concentration in Respirable Particles." *Science,* Vol. CLXXXIII (January 18, 1974), pp. 202–4.

Nef, John U. *The Conquest of the Material World.* Chicago: University of Chicago Press, 1964.

Nelson, Norton, Theodore C. Byerly, Albert C. Kolbye, Jr., *et al.* "Hazards of Mercury." *Environmental Research,* Vol. IV (1971), pp. 1–69.

Nelson, W. C., M. H. Lykins, J. Mackey, *et al.* "Mortality Among Orchard Workers Exposed to Lead Arsenate Spray: A Cohort Study." *Journal of Chronic Diseases,* Vol. XXVI (1973), pp. 105–18.

Newberne, Paul M. "Mercury in Fish: A Literature Review." *CRC Critical Reviews in Food Technology,* Vol. IV (1974), pp. 311–35.

Nilsson, Robert. *Aspects on the Toxicity of Cadmium and Its Compounds: A Review* [Ecological Research Committee Bulletin No. 7]. Stockholm: Swedish Natural Science Research Council, March, 1970.

Nova, James J., Audrey H. Nova, Robert J. Sommerville, *et al.* "Maternal Exposure to Potential Teratogens." *Journal of the American Medical Association,* Vol. CCII (December 18, 1967), pp. 1065 ff.

Novick, Sheldon. *The Careless Atom.* New York: Dell, 1969.

———. "A New Pollution Problem." *Environment,* Vol. XI (May, 1969), pp. 2–9.

———. "Toward a Nuclear Power Precipice." *Environment*, Vol. XV (March, 1973), pp. 32–40.

NSF/NASA SOLAR ENERGY PANEL, eds. *An Assessment of Solar Energy as a National Energy Resource.* College Park: University of Maryland, Department of Mechanical Engineering, December, 1972.

O'DELL, BOYD L. "Dietary Interactions of Copper and Zinc." In Delbert D. Hemphill, ed. *Proceedings University of Missouri's 1st Annual Conference on Trace Substances in Environmental Health.* Columbia: University of Missouri Press, 1968.

ODUM, HOWARD T. *Environment, Power and Society.* New York: John Wiley & Sons, 1971.

OGBURN, CHARLTON, JR. "Why the Global Income Gap Grows Wider." *Population Bulletin*, Vol. XXVI (June, 1970), pp. 3–35.

Oil, Chemical and Atomic Workers International Union. *Hazards in the Industrial Environment* [A Conference Sponsored by District 3 Council, OCAW, Atlanta, Georgia, January 23–25, 1970]. Washington, D.C.: Oil, Chemical and Atomic Workers International Union, 1970.

———. *Hazards in the Industrial Environment* [A Conference Sponsored by District 7 Council, OCAW, Fort Wayne, Indiana, October 24–26, 1969]. Washington, D.C.: Oil, Chemical and Atomic Workers International Union, 1970.

———. *Hazards in the Industrial Environment* [A Conference Sponsored by District 4 Council, OCAW, Houston, Texas, February 20–21, 1970]. Washington, D.C.: Oil, Chemical and Atomic Workers International Union, 1970.

———. *Hazards in the Industrial Environment* [A Conference Sponsored by District 8 Council, OCAW, Kenilworth, New Jersey, March 29, 1969]. Washington, D.C.: Oil, Chemical and Atomic Workers International Union, 1970.

———. *Hazards in the Industrial Environment* [A Conference Sponsored by District 5 Council, OCAW, Tulsa, Oklahoma, November 1–2, 1969]. Washington, D.C.: Oil, Chemical and Atomic Workers International Union, 1970.

OLSON, MANCUR, and HANS H. LANDSBERG. *The No-Growth Society.* New York: Norton, 1973.

O'TOOLE, THOMAS. "Environmentalists Urge Reactor Cooling Tests." *Washington Post*, February 2, 1972, p. A-2.

———. "Taiwan Getting Atomic Reactor." *Washington Post*, July 20, 1972, p. A-9.

PAGE, JOSEPH A., MARY-WIN O'BRIEN, and GARY SELLERS. *Occupational Epidemic: The Ralph Nader Task Force Report on Job Health and Safety.* Washington, D.C.: Center for Study of Responsive Law, 1972.

PARIZEK, J. "The Peculiar Toxicity of Cadmium During Pregnancy—An

Experimental 'Toxaemia of Pregnancy' Induced by Cadmium Salts."
Journal of Reproduction and Fertility, Vol. IX (1965), pp. 111–12.
———. Untitled communication. *Nature,* Vol. CLXXVIII (June 2, 1956),
p. 1037.
———, and Z. ZAHOR. "Effects of Cadmium Salts on Testicular Tissue."
Nature, Vol. CLXXVII (June 2, 1956), p. 1036.
PASSOW, H., A. ROTHSTEIN, and T. W. CLARKSON. "The General Phar-
macology of the Heavy Metals." *Pharmacological Review,* Vol. XIII (June,
1961), pp. 203, 208.
PATERSON, J. C. "Studies on the Toxicity of Inhaled Cadmium. III. The
Pathology of Cadmium Smoke Poisoning in Man and in Experimental
Animals." *Journal of Industrial Hygiene and Toxicology,* Vol. XXIX (Sep-
tember, 1947), pp. 294–95.
PATTERSON, CLAIRE C. "Lead in the Environment." *Connecticut Medicine,*
Vol. XXXV (June, 1971), pp. 347–52.
PEIRCE, CHARLES SANDERS. "The Fixation of Belief." *Popular Science Month-
ly,* Vol. XII (1877), pp. 1–15; reprinted with emendations in Philip
Weiner, ed. *Charles S. Peirce, Selected Writings: Values in a Universe of
Chance.* New York: Doubleday, 1958; Dover paperback ed. 1966.
PERLSTEIN, MEYER A., and RAMZY ATTALA. "Neurologic Sequelae of Plum-
bism in Children." *Clinical Pediatrics,* Vol. V (May, 1966), pp. 292–98.
PERRY, H. MITCHELL, JR. "Trace Elements Related to Cardiovascular Dis-
ease." *Geological Society of America Memoir 123,* 1971, pp. 171–95.
PLATT, JOHN. "What We Must Do." *Science,* Vol. CLXVI (November 28,
1969), pp. 1115–21.
PLUNKETT, E. R. *Handbook of Industrial Toxicology.* New York: Chemical
Publishing Co., 1966.
PORIES, WALTER J. "Health Effects of Trace Substances: Introductory Re-
marks." In Delbert D. Hemphill, ed. *Trace Substances in Environmental
Health—III.* Columbia: University of Missouri Press, 1970.
———, WILLIAM H. STRAIN, JENG M. HSU, and RAYMOND L. WOOSLEY, eds.
Clinical Application of Zinc Metabolism. Springfield, Ill.: Charles C.
Thomas, 1974.
———, WILLIAM H. STRAIN, CHARLES G. ROB, *et al.* "Trace Elements and
Wound Healing." In Delbert D. Hemphill, ed. *Proceedings University of
Missouri's 1st Annual Conference on Trace Substances in Environmental
Health.* Columbia: University of Missouri Press, 1968.
POTTS, C. L. "Cadmium Proteinuria." *Annals of Occupational Hygiene,* Vol.
VIII (1965), pp. 55–61.
President's Science Advisory Committee. *Restoring the Quality of Our Envi-
ronment: Report of the Environmental Pollution Panel of the President's Science
Advisory Committee.* Washington, D.C.: U.S. Government Printing Office,
1965.

Privacy Journal. Monthly from Privacy Journal (Robert Ellis Smith, ed.), P.O. Box 8844, Washington, D.C. 20003.

PROPPING, P., and F. VOGEL. "Considerations of Environment-Induced Mutations in Man." *Triangle,* Vol. XII (1973), pp. 49–55.

RABINOWITZ, M., and G. W. WETHERILL. "Identifying Sources of Lead Contamination by Stable Isotope Techniques." Typescript, 1973. [Authors are at Department of Planetary and Space Science, University of California, Los Angeles.]

RAMEL, CLAES. "Genetic Effects of Organic Mercury Compounds. I. Cytological Investigations on Allium Roots." *Hereditas,* Vol. LVII (1969), pp. 208–30.

————, and JAN MAGNUSSON. "Genetic Effects of Organic Mercury Compounds. II. Chromosome Segregation in Drosophila Melanogaster." *Hereditas,* Vol. LVII (1969), pp. 231–54.

RASMUSSEN, WAYNE D., ed. *Readings in the History of American Agriculture.* Urbana: University of Illinois Press, 1960.

REED, ROY. "Dangerous Levels of Mercury Found in Lakes and Streams in 14 Eastern States by U.S. Investigators." New York *Times,* July 9, 1970, p. 24.

REILLY, WILLIAM, ed. *The Use of Land: A Citizens' Policy Guide to Urban Growth.* New York: Crowell, 1973.

RIBICOFF, ABRAHAM. "Opening Statement of the Chairman." In United States Senate. *Chemicals and the Future of Man: Hearings Before the Subcomittee on Executive Reorganization and Government Research of the Committee on Government Operations, United States Senate, Ninety-second Congress, First Session, April 6 and 7, 1971.* Washington, D.C.: U.S. Government Printing Office, 1971.

RIDKER, RONALD G. "Population and Natural Resource Adequacy." *Resources,* No. 46 (June, 1974), p. 9.

RIPLEY, ANTHONY. "Major Safety Devices Untested at U.S. Nuclear Power Plants." New York *Times,* December 11, 1971, pp. 1, 54.

ROCKEFELLER, JOHN D., 3d, *et al. Population and the American Future.* New York: New American Library, 1972.

ROSENBLATT, ALFRED I. "Energy Crisis Spurs Development of Photovoltaic Power Sources." *Electronics,* Vol. XLVII (April 4, 1974), pp. 99–111.

ROYSTON, G. RIDDELL. "Medical Hazards of Beryllium." *The Practitioner,* Vol. CLXIV (May, 1950), pp. 391–96.

RYCKMAN, D. W., A. P. GREENSFELDER, S. JANKOVIC, and C. LUE-HING. "Recovery, Characterization, and Identification of Organic Micropollutants in Water." In Delbert D. Hemphill, ed. *Proceedings University of Missouri's 1st Annual Conference on Trace Substances in Environmental Health.* Columbia: University of Missouri Press, 1968.

RYTHER, JOHN H. "Is the World's Oxygen Supply Threatened?" *Nature,* Vol. CCXXVII (July 25, 1970), pp. 374–75.

————. "Photosynthesis and Fish Production in the Sea." *Science,* Vol. CLXVI (October 3, 1969), pp. 16–25.

SACKMAN, HAROLD, AND H. BORKO, eds. *Computers and the Problems of Society.* Montvale, N.J.: AFIPS Press [210 Summit Avenue., Montvale, N.J. 07645], 1972.

SAHA, J. G. "Significance of Mercury in the Environment." *Residue Reviews,* Vol. XLII (1972), pp. 103–63.

SALANT, WILLIAM. "The Pharmacology of Heavy Metals," *Journal of Industrial Hygiene,* Vol. II (June, 1920), pp. 72–78.

SAX, JOSEPH. *Defending the Environment: A Strategy for Citizen Action.* New York: Knopf, 1971.

SAX, N. IRVING, ed. *Dangerous Properties of Industrial Materials.* 3d ed. New York: Van Nostrand Reinhold, 1968.

SAYRE, JAMES W., EVAN CHARNEY, JAROSLAV VOSTAL and I. BARRY PLESS. "House and Hand Dust as a Potential Source of Childhood Lead Exposure." *American Journal of Diseases of Children,* Vol. CXXVII (February, 1974), pp. 167–70.

SCANLON, JOHN. "Human Fetal Hazards from Environmental Pollution with Certain Non-Essential Trace Elements." *Clinical Pediatrics,* Vol. II (March, 1972), pp. 135–41.

SCHLAEPFER, WILLIAM W. "Sequential Study of Endothelial Changes in Acute Cadmium Intoxication." *Laboratory Investigation,* Vol. XXV (1971), pp. 556–64.

SCHNEIDER, STEPHEN H., and ROGER D. DENNETT. "Climatic Barriers to Long-Term Energy Growth." *AMBIO* [A publication of the Royal Swedish Academy of Sciences], Vol. IV (1975), pp. 1–10.

SCHROEDER, HENRY A. "Airborne Metals." *Scientist and Citizen,* Vol. X (April, 1968), pp. 83–88.

————. "Cadmium." In Bureau of Mines. *Mineral Facts and Problems* [Bureau of Mines Bulletin 630]. Washington, D.C.: U.S. Government Printing Office, 1965.

————. "Cadmium as a Factor in Hypertension." *Journal of Chronic Diseases,* Vol. XVIII (1965), pp. 647–56.

————. "Editorial: The Biological Trace Elements, or Peripatetics Through the Periodic Table." *Journal of Chronic Diseases,* Vol. XVIII (1965), pp. 217–28.

————. "Inorganic Metabolism." In Bernard L. Oser, ed. *Hawk's Physiological Chemistry.* 14th ed. New York: McGraw-Hill, 1965.

————. "Metals in the Air." *Environment,* Vol. XIII (October, 1971), pp. 18–24, 29–32.

————. *Pollution, Profits & Progress.* Brattleboro, Vt.: Stephen Greene Press, 1971.

————. "Possible Relationships Between Trace Metals and Chronic Diseases." In MARVIN L. SEVEN and L. AUDREY JOHNSON, eds. *Metal Binding in Medicine.* Philadelphia: Lippincott, 1960.

―――. "Relations Between Hardness of Water and Death Rates from Certain Chronic and Degenerative Diseases in the United States." *Journal of Chronic Diseases,* Vol. XII (December, 1960), pp. 586–91.

―――. "Some Prospects for Research on Biologically Active Trace Elements." In Delbert D. Hemphill, ed. *Proceedings University of Missouri's 1st Annual Conference on Trace Elements in Environmental Health.* Columbia: University of Missouri Press, 1968.

―――. *The Trace Elements and Man: Some Positive and Negative Aspects.* Old Greenwich, Conn.: Devin-Adair Co., 1973.

―――. "Trace Elements in Degenerative Cardiovascular Disease." In Hadley L. Conn, Jr. and Orville Horowitz, eds. *Cardiac and Vascular Diseases.* Vol. II. Philadelphia: Lea & Febiger, 1971.

―――. "Trace Metals and Chronic Diseases." *Advances in Internal Medicine,* Vol. VIII (1956), pp. 259–303.

―――, and JOSEPH J. BALASSA. "Abnormal Metals in Man: Zirconium." *Journal of Chronic Diseases,* Vol. XIX (1966), pp. 573–86.

―――. "Cadmium: Uptake by Vegetables from Superphosphate in Soil." *Science,* Vol. 140 (May 17, 1963), pp. 819–20.

―――, and JEAN C. HOGENCAMP. "Abnormal Trace Metals in Man: Cadmium." *Journal of Chronic Diseases,* Vol. XIV (August, 1961), pp. 236–58.

―――, JOSEPH J. BALASSA, and ISABEL H. TIPTON. "Abnormal Trace Metals in Man—Vanadium," *Journal of Chronic Diseases,* Vol. XVI (1963), pp. 1047–71.

―――. "Abnormal Trace Elements in Man—Chromium." *Journal of Chronic Diseases,* Vol. XV (1962), p. 941.

―――. "Abnormal Trace Metals in Man—Nickel." *Journal of Chronic Diseases,* Vol. XV (1962), pp. 51–65.

―――, ALEXIS P. NASON, ISABEL H. TIPTON, and JOSEPH J. BALASSA. "Essential Trace Metals in Man: Copper." *Journal of Chronic Diseases,* Vol. XIX (1966), pp. 1004–34.

―――. "Essential Trace Metals in Man: Zinc. Relation to Environmental Cadmium." *Journal of Chronic Diseases,* Vol. XX (1967), pp. 179–210.

SCHUCK, E. A. "Discussion: Lead in Soils and Plants; Its Relationship to Traffic Volume and Proximity to Highways." *Environmental Science and Technology,* Vol. IV (March, 1970), p. 238.

SCHUTTE, KARL H. *The Biology of the Trace Elements: Their Role in Nutrition.* Philadelphia: Lippincott, 1964.

SCOTT, RACHEL. *Muscle and Blood.* New York: Dutton, 1974.

SEAMONS, ROBERT C., JR., et al., eds. *U.S. Energy Prospects: An Engineering Viewpoint.* Washington, D.C.: National Academy of Engineering, 1974.

SELIKOFF, IRVING J., and E. CUYLER HAMMOND. "III. Community Effects of

Nonoccupational Environmental Asbestos Exposure." *American Journal of Public Health,* Vol. LVIII (September, 1968), pp. 1658–66.

SELIKOFF, IRVING J., WILLIAM J. NICHOLSON, and ARTHUR M. LANGER. "Asbestos Air Pollution." *Archives of Environmental Health,* Vol. XXV (July, 1972), pp. 1–13.

SHAW, W. H. R. "Studies in Biogeochemistry—II." *Geochimica et Cosmochimica Acta,* Vol. XIX (1960), p. 210.

SINGER, S. FRED. "Human Energy Production as a Process in the Biosphere." In Dennis Flanagan, ed. *The Biosphere.* San Francisco: W. H. Freeman, 1970.

SKERFVING, STAFFAN, KERSTIN HANSSON, and JAN LINDSTEN. "Chromosome Breakage in Humans Exposed to Methyl Mercury Through Fish Consumption." *Archives of Environmental Health,* Vol. XXI (August, 1970), pp. 133–39.

SMART, N. A. "Use and Residues of Mercury Compounds in Agriculture." *Residue Reviews,* Vol. XXIII (1968), pp. 2–36.

SMITH, W. S., and C. W. GRUBER. *Atmospheric Emissions from Coal Combustion—An Inventory Guide* [National Air Pollution Control Administration Publication No. AP-24]. Cincinnati: U.S. Department of Health, Education, and Welfare, Public Health Service, Division of Air Pollution, April, 1966.

SOCOLOW, ROBERT, MARK ROSS, D. L. HARTLEY, S. D. SILVERSTEIN, and S. M. BERMAN. "Technical Aspects of Efficient Energy Utilization; Summer Study 1974, American Physical Society, Summary." Typescript, n.p., n.d. [Dr. Socolow is with the Center for Environmental Studies, Princeton University.]

SPANGLER, WILLIAM J., JAMES L. SPIGARELLI, JOSEPH M. ROSE, and HOPE M. MILLER. "Methylmercury: Bacterial Degradation in Lake Sediments." *Science,* Vol. CLXXX (April 13, 1973), pp. 192–93.

SPEAR, PAUL W. "Fight Against Lead Poisoning" [a letter to the editor]. New York *Times,* July 1, 1971, p. 46.

SPEAR, ROBERT C., and EDDIE WEI. "Methylmercury Toxicity: A Probabilistic Assessment." In Donald R. Buhler, ed. *Mercury in the Western Environment.* Corvallis, Oregon: Continuing Education Publications [Ext. Hall Annex, Corvallis, Oregon 97331], 1973.

SPEIL, S., and J. P. LEINEWEBER. "Asbestos Minerals in Modern Technology." *Environmental Research,* Vol. II (1969), pp. 166–208.

SPYKER, JOAN M., SHELDON B. SPARBER, and ALAN M. GOLDBERG. "Subtle Consequences of Methylmercury Exposure: Behavioral Deviations in Offspring of Treated Mothers." *Science,* Vol. CLXXVII (August 18, 1972), pp. 621–23.

SROLE, LEO, et al. *Mental Health in the Metropolis: The Midtown Manhattan Study.* Vol. I. New York: McGraw-Hill, 1962.

STARR, CHAUNCEY. "Energy and Power." In Board of Editors of *Scientific American,* eds. *Energy and Power.* San Francisco: W. H. Freeman, 1971.

STERNGLASS, ERNEST J. *Low-Level Radiation.* New York: Ballantine, 1972.

STUDY OF CRITICAL ENVIRONMENTAL PROBLEMS. *Man's Impact on the Global Environment.* Cambridge, Mass.: MIT Press, 1970.

STUPKA, GARY, *et al.* "An Analysis of the Behavior and Effects of Lead Pollution on a Terrestrial and Aquatic Ecosystem in the Upper Great Lakes Region. Mimeographed. De Kalb: Northern Illinois University, December, 1971.

SUMMERS, CLAUDE M. "The Conversion of Energy." In Board of Editors of *Scientific American,* eds. *Energy and Power.* San Francisco: W. H. Freeman, 1971.

SWENERTON, HELENE, and LUCILLE S. HURLEY. "Teratogenic Effects of a Chelating Agent and Their Prevention by Zinc." *Science,* Vol. CLXXIII (July 2, 1971), pp. 62–64.

TAYLOR, AVERY. "Warning: Your Wall May Be a Killer." *Environmental Action,* Vol. III (June 26, 1971), p. 10.

TAYLOR, THEODORE B., and CHARLES C. HUMPSTONE. *The Restoration of the Earth.* New York: Harper & Row, 1973.

THOMPSON, PEGGY. "New Findings Show That Lead Is An Everywhere Poison." *National Observer,* June 22, 1974, pp. 1, 12.

THUROW, LESTER C., and ROBERT E. B. LUCAS. *The American Distribution of Income: A Structural Problem; A Study Prepared for the Use of the Joint Economic Committee, Congress of the United States.* Washington, D.C.: U.S. Government Printing Office, 1972.

TIPTON, ISABEL H. "Geographic Variations in Concentrations of Trace Elements in Human Tissue." Typescript ["Research jointly sponsored by the U.S. Atomic Energy Commission under contract with the Union Carbide Corporation and the University of Tennessee, through subcontract No. 1495 under W7407 eng 26," n.d. (1964)].

TOWNSHEND, R. H. "A Case of Acute Cadmium Pneumonitis: Lung Function Tests During a Four-Year Follow-Up." *British Journal of Industrial Medicine,* Vol. XXV (1968), pp. 68–71.

TSUCHIYA, KENZABURO. "Causation of Ouch-Ouch Disease [Itai-Itai Byo], Part I: Nature of the Disease." *Keio Journal of Medicine,* Vol. XVIII (1969), pp. 181–94.

————. "Causation of Ouch-Ouch Disease, Part II: Epidemiology and Evaluation." *Keio Journal of Medicine,* Vol. XVIII (1969), pp. 195–211.

————. "Proteinuria of Workers Exposed to Cadmium Fume." *Archives of Environmental Health,* Vol. XIV (June, 1967), pp. 875–80.

————, and SUSUMU HARASHIMA. "Lead Exposure and the Derivation of Maximum Allowable Concentrations and Threshold Limit Values." *British Journal of Industrial Medicine,* Vol. XXII (1965), pp. 181–86.

TUCKER, ANTHONY. *Toxic Metals.* New York: Ballantine, 1972.

UCHIDA, M., K. HIRAKAWA, and T. INOUE, "Biochemical Studies on Minamata Disease, IV. Isolation and Chemical Identification of the Mercury Compound in the Toxic Shellfish with Special Reference to the Causal Agent of the Disease—Organic Mercury, Methylmercuric Sulfide." *Kumamoto Medical Journal,* Vol. XIV (1961), pp. 181–87.

UETZ, GEORGE, and DONALD LEE JOHNSON. "Breaking the Web." *Environment,* Vol. XVI (December, 1974), pp. 31–39.

ULLMAN, WILLIAM K. "Lead in the Connecticut Environment." *Connecticut Medicine,* Vol. XXXV (June, 1971), pp. 360–62.

ULMER, DAVID L., and BERT L. VALLEE. "Effects of Lead on Biochemical Systems." In Delbert D. Hemphill, ed. *Trace Substances in Environmental Health—II.* Columbia: University of Missouri Press, 1969.

UNDERWOOD, E. J. *Trace Elements in Human and Animal Nutrition.* 3d ed. New York: Academic Press, 1971.

United Press International. "Chromosome Damage, Mercury Found in Remote South American Indians." Albuquerque *Tribune,* January 2, 1974, p. G-8.

———. "Lead Hurt Children; Smelter Is Blamed." Albuquerque *Journal,* May 11, 1974, p. B-12.

———. "Retardation Panel Cites Environment." Washington *Post,* January 22, 1972, p. A-2.

United States Bureau of the Census. *Statistical Abstract of the United States 1973.* 94th ed. Washington, D.C.: U.S. Government Printing Office, 1973.

United States Department of Agriculture. *Changes in Farm Production and Efficiency: A Summary Report, 1970* [USDA Statistical Bulletin No. 233]. Washington, D.C.: U.S. Government Printing Office, 1970.

United States Department of Health, Education, and Welfare. *Criteria for a Recommended Standard . . . Occupational Exposure to Asbestos* [HSM 72-10267] . Washington, D.C.: U.S. Department of Health, Education, and Welfare, Public Health Service, Health Services and Mental Health Administration, National Institute for Occupational Safety and Health, 1972.

———. *Diabetes Mellitus Mortality in the United States—1950–1967* [Public Health Service Publication No. 1000, Series 20, No. 10]. Washington, D.C.: U.S. Government Printing Office, July, 1971.

———. *Facts of Life and Death* [Public Health Service Publication No. 600, revised 1970]. Washington, D.C.: U.S. Government Printing Office, 1970.

———. "News Release for Release February, 13, 1970" [No. 70-2A]. Washington, D.C.: Department of Health, Education, and Welfare, 1970.

————. "News Release for Release Feb. 5, 1971." *HEW News.* Washington, D.C.: Department of Health, Education, and Welfare, Public Health Service, Food and Drug Administration, 1971.

————. "News Release for Release February 8, 1971" [No. 71-6]. Washington, D.C.: Department of Health, Education, and Welfare, 1971.

————. "News Release for Release July 13, 1971" [No. 71-43]. Washington, D.C.: Department of Health, Education, and Welfare, 1971.

————. *"Public Health Service Drinking Water Standards 1962"* [Public Health Service Publication No. 956]. Washington, D.C.: U.S. Government Printing Office, 1969.

United States Department of the Interior. "Secretary Hickel Reports Mercury Discharges Reduced by 86 Percent." News release ["For Immediate Release Sept. 16, 1970"]. Washington, D.C.: U.S. Department of the Interior, 1970.

United States Food and Drug Administration. "News Release for Release March 26, 1971" [No. 71-15]. Washington, D.C.: Food and Drug Administration, 1971.

United States Geological Survey. *Mercury in the Environment* [Geological Survey Professional Paper 713]. Washington, D.C.: U.S. Government Printing Office, 1970.

————. *Reconnaissance of Selected Minor Elements in Surface Waters of the United States—October, 1970* [Geological Survey Circular 643]. Washington, D.C.: U.S. Government Printing Office, 1970.

United States Senate. *Chemicals and the Future of Man: Hearings Before the Subcommittee on Executive Reorganization and Government Research of the Committee on Government Operations, United States Senate, Ninety-second Congress, First Session, April 6 and 7, 1971.* Washington, D.C.: U.S. Government Printing Office, 1971.

————. *Parity Returns Position of Farmers: Report to the Congress of the United States Pursuant to Section 705 of the Food and Agriculture Act of 1965 by the Department of Agriculture, Ninetieth Congress, First Session* [Senate Document No. 44, August 10, 1967]. Washington, D.C.: U.S. Government Printing Office, 1967.

Van Dyne, George M., ed. *The Ecosystem Concept in Resource Management.* New York: Academic Press, 1969.

Viets, Frank G., Jr. "Zinc Deficiency in the Soil-Plant System." In Ananda Shiva Prasad, ed. *Zinc Metabolism.* Springfield, Ill.: Charles C. Thomas, 1966.

Vitamin Information Bureau. "Zinc Backgrounder." Mimeographed. New York: Vitamin Information Bureau [575 Lexington Avenue, New York, N.Y. 10017], n.d.

Voors, A. W., M. S. Shuman, G. P. Woodward, and P. N. Gallager. "Arterial Lead Levels and Cardiac Death: A Hypothesis." *Environmental Health Perspectives,* June, 1973, p. 97.

WADE, NICHOLAS. "Raw Materials: U.S. Grows More Vulnerable to Third World Cartels." *Science*, Vol. CLXXXIII (January 18, 1974), pp. 185–86.

WADLEIGH, CECIL H. *Wastes in Relation to Agriculture and Forestry* [USDA Miscellaneous Publication No. 1065]. Washington, D.C.: U.S. Government Printing Office, 1968, 1972.

WAGNER, J. C. "Guest Editorial—Asbestos Cancers." *Journal of the National Cancer Institute*, Vol. XLVI (May, 1971), p. v.

WAGNER, RICHARD H. *Environment and Man*. New York: W. W. Norton, 1971.

WALDRON, MARTIN. "Mercury in Food: A Family Tragedy." *New York Times*, August 10, 1970, p. A-1.

WALLACE, ROBIN A., WILLIAM FULKERSON, WILBUR D. SHULTS, and WILLIAM S. LYON. *Mercury in the Environment: The Human Element* [ORNL NSF-EP-1]. Oak Ridge, Tenn.: Oak Ridge National Laboratory, January, 1971.

WALLICK, FRANK. *The American Worker: An Endangered Species*. New York: Ballantine, 1972.

WARREN, HARRY V., ROBERT E. DELAVAULT, and CHRISTINE H. CROSS. "Base Metal Pollution in Soils." In Delbert D. Hemphill, ed. *Trace Substances in Environmental Health—III*. Columbia: University of Missouri Press, 1970.

WEINBERG, ALVIN M. "The Moral Imperatives of Nuclear Energy." *Nuclear News*, December, 1971, pp. 33–37.

WELLFORD, HARRISON. *Sowing the Wind*. New York: Grossman, 1972.

WESTOO, GUNNEL. "Methylmercury as Percentage of Total Mercury in Flesh and Viscera of Salmon and Sea Trout of Various Ages." *Science*, Vol. CLXXI (August 10, 1973), pp. 567–68.

———. "Methylmercury Compounds in Animal Foods." In Morton W. Miller and George G. Berg, eds. *Chemical Fallout: Current Research on Persistent Pesticides*. Springfield, Ill.: Charles C. Thomas, 1969.

———. "Methylmercury Compounds in Fish." *Oikos, Acta Oecologica Scandinavica* [Supp. 9] (1967), pp. 11–12.

WILLIAMS, ROGER J. *Biochemical Individuality: The Basis for the Genetotrophic Concept*. New York: John Wiley & Sons, 1956.

———. *Nutrition Against Disease*. New York: Pitman, 1971.

———. *Nutrition in a Nutshell*. Garden City, N.Y.: Doubleday, 1962.

———, E. BEERSTECHER, JR., and L. J. BERRY. "The Concept of Genetotrophic Disease." *The Lancet*, February 18, 1950, pp. 287–89.

WILSON, PETERS D. "Massive Soil Losses Predicted in Crop Conversion." *Conservation News*, Vol. XXXIX (April 1, 1974), pp. 2–4.

WISE, WILLIAM. *Killer Smog*. New York: Ballantine, 1970.

WOOD, J. M., F. SCOTT KENNEDY, and C. G. ROSEN. "Synthesis of Methylmercury Compounds by Extracts of a Methanogenic Bacterium." *Nature*, Vol. 220 (October 12, 1968), pp. 173–74.

Woodwell, George M. "The Energy Cycle of the Biosphere." In Dennis Flanagan, ed. *The Biosphere*. San Francisco: W. H. Freeman, 1970.

Workbook, The. Monthly from Southwest Research and Information Center, P.O. Box 4524, Albuquerque, N.M. 87106.

Wright, George F. "Exposure to Alkyl Mercury." *Science*, Vol. CLXXIV (November 19, 1971), p. 771.

Zielinski, John F. *Analyses of Factors in Beryllium Associated Diseases*. Cleveland: Brush Beryllium Company [5209 Euclid Avenue, Cleveland, Ohio 44103], 1962.

Index

Note: The letter "t" following a page number indicates a table.